AF411495

Palgrave Studies in Young People and Politics

Series Editors
James Sloam
Department of Politics and International Relations
Royal Holloway, University of London
Egham, UK

Constance Flanagan
School of Human Ecology
University of Wisconsin–Madison
Madison, WI, USA

Bronwyn Hayward
School of Social and Political Sciences
University of Canterbury
Christchurch, New Zealand

Over the past few decades, many democracies have experienced low or falling voter turnout and a sharp decline in the members of mainstream political parties. These trends are most striking amongst young people, who have become alienated from mainstream electoral politics in many countries across the world. Young people are today faced by a particularly tough environment. From worsening levels of child poverty, to large increases in youth unemployment, to cuts in youth services and education budgets, public policy responses to the financial crisis have placed a disproportionate burden on the young.

This book series will provide an in-depth investigation of the changing nature of youth civic and political engagement. We particularly welcome contributions looking at:

- Youth political participation: for example, voting, demonstrations, and consumer politics
- The engagement of young people in civic and political institutions, such as political parties, NGOs and new social movements
- The influence of technology, the news media and social media on young people's politics
- How democratic innovations, such as social institutions, electoral reform, civic education, can rejuvenate democracy
- The civic and political development of young people during their transition from childhood to adulthood (political socialisation)
- Young people's diverse civic and political identities, as defined by issues of gender, class and ethnicity
- Key themes in public policy affecting younger citizens—e.g. youth (un)employment and education
- Cross-cutting themes such as intergenerational inequality, social mobility, and participation in policy-making—e.g. school councils, youth parliaments and youth wings of political parties

The series will incorporate a mixture of pivot publications (25,000–50,000 words), full-length monographs and edited volumes that will analyse these issues within individual countries, comparatively, and/ or through the lenses of different case studies.

More information about this series at
http://www.palgrave.com/gp/series/15478

Billur Aslan Ozgul

Leading Protests in the Digital Age

Youth Activism in Egypt and Syria

Billur Aslan Ozgul
Department of Social and Political
Sciences
Brunel University London
London, UK

Palgrave Studies in Young People and Politics
ISBN 978-3-030-25449-0   ISBN 978-3-030-25450-6   (eBook)
https://doi.org/10.1007/978-3-030-25450-6

Cover credit: Joel Carillet/Getty Images

This Palgrave Macmillan imprint is published by the registered company Springer Nature Switzerland AG
The registered company address is: Gewerbestrasse 11, 6330 Cham, Switzerland

To Ahmet, thanks for making this journey even more memorable

Acknowledgements

This book has benefited from the valuable ideas and advice of many people. First of all, I would like to thank my Egyptian and Syrian interviewees, who spent hours answering my questions and shared with me their experiences, hopes and sorrows. I owe a special debt of thanks to Abdelrahman Ayyash, Abed Al Sarraj, Amer Omar, Bilal Zaiter, Caroline Ayyoub, Islam Talaat, Iyas Kaadouni, Radamis Hany Zaky, Sherif Azer and Sherif Alaa for their contributions and for introducing me to many great revolutionaries. This manuscript would not have been possible without them. I would also add that the support of Mina Hany, his guidance in Cairo and his translations were of crucial importance for the preparation of this manuscript. In the summer of 2013, I spent hours at the Syrian National Coalition Media office's employees, listening to their protest experiences in different parts of Syria; I warmly thank them for allowing me to visit their offices and conduct interviews with them. In addition, I am tremendously grateful for the time of Abdelrahman Mansour, Ahmed Samih, Enrico Angelis, Esraa Abdel Fattah, Mahmoud Masri (pseudonym), Wael Abbas and Wael Eskandar. This book greatly benefitted from the recounting of their experiences and knowledge.

I would also like to thank the editorial team at Palgrave Macmillan, and Ambra Finotello and Anne-Kathrin Birchley-Brun, who have been very supportive in the writing stage and patiently answered my questions. I am also most grateful to James Sloam for his encouragement, support and valuable advice on my draft chapters.

I began this study when I was a Ph.D. candidate at the New Political Communication Unit at Royal Holloway, University of London, and continued at the Centre of Media and Politics at Bournemouth University. Both institutions provided me with many opportunities to conduct the study, for which I am tremendously grateful.

There are also many colleagues who were inspiring and substantially supported this study with their intellectual assistance. I owe a special thanks to Ben O'Loughlin, who has always been an inspiration for me. His valuable advice, support and enthusiasm enabled this manuscript to be drafted and progress to its best potential. I also very much appreciate the useful feedback and constructive criticism of Juan Cole, Cristian Vaccari, Andrew Chadwick and Oliver Heath; their theoretical and methodological advice greatly improved this book. At the beginning of this research, Alister Miskimmon created an incredible opportunity for me by introducing the DGAP conference that took place in Cairo, and this is where I first met with Egyptian activists, which was a turning point in my life and for which I am extremely grateful. I am also very fortunate to have been surrounded by excellent research mentors, particularly Dan Jackson and Darren Lilleker, who contributed much my thinking while this book was being written.

I gratefully acknowledge that some of the materials presented in Chapters 5 and 6 appear in Aslan (2015). The Mobilisation Process of Syria's Activists: The Symbiotic Relationship Between the Use of Information and Communication Technologies and the Political Culture, *International Journal of Communication* 9(19): 2507–2525. The manuscript has also been greatly embellished from the proofreading of Christopher Wood. I very much appreciate his assistance and friendship.

None of this would have been possible without the support of my family. I would like to thank my mother, Neslihan, who has always encouraged me to follow my dreams and taught me to challenge all types of gender stereotypes. To my father, Suleyman, who is the main source of my interest in politics and political participation, to my brother, Metin, for always comforting me with his presence, and to my grandparents whom I love dearly. Finally, I dedicate this book to my husband, Ahmet, who sustained me while conducting this research and writing the manuscript over seven years. I am thankful for his love, encouragement, support and assistance.

Contents

Abbreviations

ABC	Australian Broadcasting Corporation
ANA	Activist News Organisation
BBC	British Broadcasting Corporation
CBC	Canadian Broadcasting Corporation
CNN	Cable News Network
GDP	Gross Domestic Product
ICTs	Information Communication Technologies
IMF	International Monetary Fund
IRI	International Republican Institute
ISP	Internet Service Provider
KKSFP	Kullena Khaled Said Facebook Page
LCC	Local Coordination Committees
NAC	National Association for Change
NDP	National Democratic Party
NGO	Non-Governmental Organisation
RYC	Revolutionary Youth Coalition
SEA	Syrian Electronic Army
SNC	Syrian National Coalition
SNCM	Syrian National Coalition Media
SRFP	Syrian Revolution Facebook Page
SRGC	Syrian Revolution General Commission
TV	Television
UK	United Kingdom
US	United States
WUNC	Worthiness, Unity, Numbers and Commitment

Introduction

On 26 January 2011, Nisrin,[1] a young Egyptian activist, and her friend were sitting in a restaurant called Mezzaluna in Zamalek, Cairo. The previous day, they had witnessed something that their generation had never seen before. Thousands of Egyptians of all ages had flooded into Cairo's Tahrir Square. The organising group had already planned to protest against police repression on 25 January, but the waves of protests coming from Tunisia brought them further hope. They raised their demands and, online and offline, started calling for an end to Mubarak's rule. Filled with the emotions of the last few days, Nisrin met with her friend to talk about how they could join the protests that would take place on Friday 28. The waitresses within the restaurant who eavesdropped on Nisrin and her friend stopped by their table and joined the conversation. Nisrin was surprised to see the attention paid to "Kullena Khaled Said" ("We are all Khaled Said") Facebook page, the most popular online platform during the Egyptian revolution of early 2011. The waitresses were not on Facebook at the time, yet they all knew about the protest call on the Facebook page and wanted to hear more about it. When the waitresses finished their shifts, no one left the table. They all debated and discussed what would happen on Friday 28. Amazed by the attention, Nisrin and her friend strongly encouraged the waitresses to join them during the protests.

On 27 January, Nisrin slept at her friend's place on Gezira Street so that they could walk to Tahrir Square, where the protesters would gather

© The Author(s) 2020

B. Aslan Ozgul, *Leading Protests in the Digital Age*,
Palgrave Studies in Young People and Politics,
https://doi.org/10.1007/978-3-030-25450-6_1

the next day. Two years later, in a café in London, Nisrin describes the scene to me with the same excitement she felt on that day.

> Gezira is close to Tahrir. On the streets, there were people who were looking like me, like American University of Cairo students. They really didn't have the courage to face the violence alone. So, we marched together. But we were subject to so many gas bombs. Everyone was tired and we decided to sit on the sidewalks. It was horrible, we were extremely exhausted. Then, all of a sudden, we saw a big group composed of young men all wearing tank tops coming and at the front of the group, there were the two waitresses from Mezzaluna. They were leading the whole group!

Nisrin's memories of 28 January reflect the enthusiasm and desire of both poor and middle-class young Egyptians for political change. The unification of the different groups on 28 January was the result of a communication strategy carefully planned by the Egyptian activists. By using both online and offline communication tactics, which constitute the topic of this book, the activists reached different segments of society and mobilised them. The day marked a turning point in Egyptian history. Coming out from Friday prayers, tens of thousands of Egyptians all over the country engaged in confrontations with police to demand the end of President Hosni Mubarak's rule (Abaza and Youssef 2011). Learning from their past experience, the organising group was not expecting a revolution similar to that in Tunisia.[2] However, the thousands of protestors on the small streets of Cairo battled the police, making demands for an end to hunger, poverty, unemployment, inflation, rising rents and the looting of public wealth (Cole 2014: 148). Attacked by the protesters, the police stations under fire signalled the beginning of the end of Mubarak's three decades of rule. Only 15 days later, Mubarak stepped down as president, handing power to the army. The wave of protests which started in Tunisia brought hope to the Egyptians and encouraged the public to pour out onto the streets. This hope then spread to Libya, Yemen, Syria and Bahrain. The protests have been described by scholars as cascading popular democracy movements (Howard and Hussein 2013: 3). The online posts, about revolution and liberty, went viral and were widely forwarded to larger circles of people until they reached saturation point (Cole 2014: 12–13).

However, by the end of 2012, the emotions of hope and joy that had once accompanied the word "revolution" in the Middle East had vanished. In the two countries upon which this book focuses, democratic rule has never taken root. Instead, Egypt has embarked on another era of military rule, while Syria has become one of the most desolate places on Earth (Alaoui 2016). Due to the authoritarian backlash, the voices of protesters were utterly drowned out in both countries, and publication of books on the mobilisation attempts of activists has dwindled. The crackdown on activists created a pressing need for those involved to voice their opinions and bring attention to the atrocities they experienced during the protests.

In this book, I aim to present a deep investigation of the protest organisation and mobilisation tactics of young activists in the digital age, and seek patterns as to which tactics worked well and which created the high risks associated with involvement in digitally supported movements. For this aim, I focus on the uprisings that took place in Egypt and Syria during 2011. Unsurprisingly, with the proliferation of digitally supported movements in 2011, many scholars have already addressed the historical origins of the Arab uprisings and the mobilisation attempts of the activists. Particularly, the role of information and communication technologies (ICTs) in protests became a popular topic in the communications field (Gerbaudo 2012: 3). This book builds on this research and analyses how activists used ICTs during the 2011 protests in Syria and Egypt. It also develops the research by offering rich and unique data on the different mobilisation techniques and protest experiences of key Egyptian and Syrian activists.

Moreover, the majority of communication studies on the Arab uprisings focus on Tunisia and Egypt, where ICTs played a crucial role (Lim 2012). The development of digital activism in Egypt, which led to the formation of ElBaradei's Facebook page and contributed to the removal from power of Hosni Mubarak's regime, was studied in detail by scholars (Faris 2013: 4). They often drew a connection between technology diffusion, the use of social media and political change (Castells 2012; Howard and Parks 2012: 360). Some also drew attention to the fact that individual members of the public are not inclined to join formal organisations for protest organisation. The Internet is used as an organisation tool, and the rapid flow of messages from one originator to potentially millions of people has replaced the role of social movement organisations, leaders and their internal and external resources in the mobilisation process behind these uprisings (Castells 2012: 15; Bennett and Segerberg 2013: 1). However, does the Internet always have a positive impact on the mobilisation process of protests in similar regimes?

If not, what were the structural factors that affect the use of ICTs? I started this research with these questions in mind. To respond to them, besides countries such as Egypt where protesters successfully adopted ICTs, research must focus on countries where ICTs could not become the main organisational hub of the protesters. So far, studies on countries where protesters struggled to mobilise via digital media channels have been somewhat anecdotal. The Syrian uprising was a great example of this problem. Despite the ongoing armed conflict in Syria, few studies have explored in detail the organisation and mobilisation attempts of the peaceful activists who engaged in demonstrations in early 2011 in Syria. The different mobilisation tactics undertaken by Syrian and Egyptian activists during the 2011 protests and the reasons behind these choices offer the perfect opportunity to study the organisation and mobilisation of digitally supported movements in different contexts.

Method

For a rigorous and accurate assessment of the role of ICTs in the protests, this book adopts a comparative approach. I argue that the comparative method is the best approach for shedding light on the diverse and shared experiences of Arab countries with ICTs. It also enables in-depth insight into different cases and captures the complexity of these (Rihoux 2006). Drawing on semi-structured interviews with activists, this book compares actors' use of different communication methods for the organisation, mobilisation, co-ordination and sustainment of the peaceful components of the protests in Syria and Egypt.

As the book investigates the differing roles of ICTs in similar regimes, it implies the most similar system design used in similar cases with different outcomes (Rihoux 2006: 685). In the most similar system design, the researcher needs to "choose objects of research systems that are as similar as possible, except with regard to the phenomenon, the effect of which we are interested in assessing" (Anckar 2008: 389). Hence, the cases need to be similar enough in terms of a large number of characteristics, but dissimilar regarding the variable between which a relationship is hypothesised (Lijphart 1975: 159). The cases of Syria and Egypt are useful comparison sets, as they are representative of the wider domain of "Arab uprisings" and share a "common set of language, similar if not shared media systems, consistently authoritarian regimes and a rapidly increasing level of technology diffusion" (Howard and Hussein 2013: 10).

These cases were specifically chosen from amongst the other Arab uprisings, as they were also different enough from one another to test the effect of these differences (Seawright and Gerring 2008: 296). While classifying the regime types of Egypt and Syria, I refer to the prominent social movement theorist and political scientist Charles Tilly's classification of regime types (2006: 21). Tilly differentiates the types of regime by comparing two dimensions of variation: (1) the level of democracy and (2) governmental capacity. The level of democracy reflects the "extent to which persons subject to the government's authority have broad, equal rights to influence governmental affairs and to receive protection from arbitrary governmental action" (Tilly 2006: 21). On the other hand, by governmental capacity, Tilly (2006) tests the degree to which governmental actions affect the distribution of populations, activities and resources within the government's jurisdiction, relative to some standard of quality and efficiency. By taking into account these two dimensions, Tilly (2006) then divides the regimes into four crude types: high-capacity non-democratic; high-capacity democratic; low-capacity non-democratic; and low-capacity democratic. When Egypt and Syria's governmental capacity and democracy level in 2010 are analysed, it can be seen that Egypt's democracy index, with regard to free and fair elections and civil liberty provision, was 3.02 out of 10. On the other hand, Syria's democracy index was 2.32 out of 10 (Economist Intelligence Unit 2010). Hence, both countries ranked as non-democratic regimes, but with different democracy levels.

Since professional freedom-tracking agencies such as Freedom House do not deal with governmental capacity directly, Tilly offers an alternative method to classify regimes according to their governmental capacity. According to Tilly's classification, high-capacity non-democratic regimes differ from others by "many prescribed performances, narrow range of tolerated performances, almost all contention by means of forbidden performances; high involvement of government agents in contention (often as principals); medium level of violence in contentious interactions" (Tilly 2006: 81). Both Egypt and Syria showed the features of high-capacity non-democratic regimes before 2011. The violent response of the Mubarak regime towards the protesters during the Second Intifada, Kefaya and Labour movements and the brutal repression of the Assad regime against the 1982 Muslim Brotherhood uprising were examples of high involvement of government agents in non-tolerated performances. Hence, according to Tilly's classification of regime types, both countries were examples of high-capacity non-democratic regime behaviour. Comparing these two cases enabled me to test whether ICTs can always replace the roles of organisations, leaders and collective identity during a movement in similar regimes.

The time periods for analysis in both cases were chosen according to the length of the peaceful protests in these countries. While the peaceful protests in Syria took place for approximately six months, in Egypt, the protesters succeeded in overthrowing President Mubarak in a mere 18 days. The different stages of peaceful uprising are explained chronologically in each of the case studies in this book. The empirical chapters trace the actors' communication methods, repertoires, development and their interaction with the state. The case studies in this research aim to provide detail on the complexities in each country that large-N quantitative studies would lack.

The case studies were enriched by semi-structured interviews with 45 activists who took part either in the organisation or mobilisation period of the protests, as well as 2 international journalists who worked in Egypt and/or Syria. I conducted these interviews in Istanbul, London and Cairo between 2011 and 2015, and eight of those were conducted via Skype. They ranged in length between 60 and 90 minutes. To shed light on online and offline activism in both countries, the interviewees included social media administrators and members of the organising groups on the ground. In Egypt, I conducted the interviews with prominent Internet figures, such as Wael Abbas, Esraa Abdel Fattah, Abdelrahman Mansour, as well as with activists who had played prominent roles in the organising groups on the ground. I also conducted interviews with young people from the Muslim Brotherhood, the April 6 movement, leftist liberal groups and human rights activists. On the other hand, I chose the Syrian activists with consideration of the ethnic and religious diversity of the Syrian opposition. They also reflected the background of the different cities that hosted peaceful protests in 2011, such as Darayya, Daraa, Damascus, Tartus, Latakia, Homs, Hama, Idlip, Saraqib and Deir ez-Zor.

Some participants' surnames are not given and ten participants identified with a pseudonym for safety reasons. The interviews were structured around 10 questions: *Did the activists have past protest experience? How did the activists hear about the protests and get involved in the collective action? What was their aim in joining the protests? How did they mobilise, co-ordinate and sustain the protests? How did they get their news during the uprising? How did they use the Internet or other communication methods, such as face-to-face communication and mobile phones, during the uprising? Do they think that the Internet can replace the role of face-to-face communication, or does it add something new? How did they find the regime's tactics in the online and offline spheres? Why is the Egyptian uprising described as more peaceful and rapid than the Syrian uprising? Do the activists still have hope for political change in their countries?*

These questions intersect with several distinct literatures in the fields of political science, communication and sociology, which I will review below. Moreover, throughout the book, I develop three themes that serve as touchstones for my investigation of the digitally supported movements: repertoires, leadership and resources. In the following discussion, I briefly introduce these themes and conclude with a more detailed overview of the book.

Online Struggle in Countries with Weak Repertoires

At the start of the Arab uprisings, the role of ICTs in protests had garnered considerable attention amongst pundits and scholars and we had seen enthusiastic celebration of their power. As the most well-known proponent of cyber optimist vision, Clay Shirky (2009) describes new media as a tool, fostering participation and increasing freedom. He argues that social tools trigger collective action by decreasing the costs of communication and removing "two old obstacles – locality of information and barriers to group reaction". In contrast, these technologies have been dismissed by a handful of scholars and journalists such as Evgeny Morozov (2012), Malcolm Gladwell (2010), Will Heaven (2009) and Jay Rosen (2011). They highlight that technologies such as social media sites have been used to reinforce dictators and threaten dissidents. Yet, as the research intensified, the abstraction and essentialism underlying this debate was criticised for analysing neither the interaction of social media with other forms of communication nor the political contexts within which these movements have emerged (Aouragh and Alexander 2011; Gerbaudo 2012). Aouragh and Alexander (2011: 1345), for instance, who conducted research on the 2011 Egyptian protests, urge scholars to leave behind the perspectives that isolate the Internet from other media. The Egyptian protests were an example of the powerful synergy between social media and satellite broadcasters. For this reason, in this book I define such movements in the digital age as "digitally supported", rather than "digitally enabled" protests.

Although ICTs did not bring about anything in and of themselves, they had numerous impacts on the movements (Cole 2014: 12). If these opportunities are analysed, we can first see that ICTs offer a means of mass communication in which the public share hope and sorrow. They connect individual members of the public to each other. In turn, this connection induces enthusiasm and motivation to protest (Castells 2012: 219). According to Castells (2012: 219), enthusiastic networked individuals can overcome fear and are transformed into conscious and collective actors. However, can networked individuals always overcome fear, connect with each other via ICTs and create this kind of unification in similar regimes with systematic differences? This question seems to shift the discussion about the opportunities that ICTs present to certain movements to the real capabilities of technology.

The interviews with Syrian activists within this book will show us that the repertoire of protesters is the main dynamic that shapes their communication capabilities and consequently their ICT use. A social movement repertoire is a term coined by Charles Tilly (2006) which refers to the variable ensemble of performances by drawing on identities, social ties and the past protest experience of protesters. These performances can include marches, rallies, processions, demonstrations, public meetings, blockades and so on (Tilly 2006: 53).

Egyptians had already formed social ties with different ideological groups during the protests and uprisings dating back to 1952. Young people with different ideologies also built virtual communities prior to the 2011 Egyptian protests.[3] In her book *Revolution in the Age of Social Media*, Herrera describes how this new generation experienced exponentially increasing rates of connectivity (Herrera 2014: 6). In contrast to the strong repertoire of Egyptian activists, there was a lack of experienced community that was capable of organising a solidly structured movement in Syria before 2011. The repertoires of Syrians, including social ties, past performances and collective identities, could be described as weak.

Tilly's (2006) theories on repertoires help us to explain this difference between the two countries. The repertoires of the activists in these two countries changed according to variations of regime type. As explained above, Egypt and Syria were in the same category of regime type, of high-capacity non-democratic regimes defined by Tilly (2006). Yet, they showed major differences in election system, diffusion of power in the regime and level of repression. As will be explained in Chapter 2, unlike

Egypt, which had a multiparty presidential system, Syria has been ruled by a one-party system that prevented any independent political parties or figures from emerging since the 1963 military coup. Unlike Egypt, where power is centralised on the president, the power of the Syrian regime was divided between the army, Ba'ath party and security forces. Yet, the power holders in Syria were linked to each other with sectarian ties. The major figures of the party and key officers of the army were from the same Alawite sect as President Al Assad. Due to these ties, the rulers of Syria applied different combinations of coercion than Egypt against opposition groups. Fearful of being pushed out of power, the Syrian army had unconditionally supported the President and prevented the appearance of any kind of protests in the country (Leverett 2005: 37). Syrians witnessed the harsh approach of the regime to any possible uprising in the country during the Hama massacre in 1982. The regime's bloody repression of the Muslim Brotherhood uprising resulted in the slaughter of over 10,000 people (Phillips 2015: 366). Incorporating one-fifth of the labour force, this military apparatus encompassed enormous resources (Alvarez-Ossorio 2012). The regime's survival was also secured by security forces with firepower. I will discuss the high level of repression employed by the state against the media and at the ground level in Chapter 2.

Semiotic practices were another medium of repression that obstructed the appearance of any opposition movement in Syria. In her book *Ambiguities of Domination*, Lisa Weeden defines these semiotic practices as the language and symbolism used by the Syrian regime (Weeden 2002: 714). They were inscribed in concrete actions and operated to produce observable effects. For instance, symbols used by the regime, such as creating a cult of personality around the president, were semiotic practices that operated as disciplinary devices in Syria. The implanting of formulaic rhetoric and self-serving state symbolism into the daily life of citizens led to reduced societal autonomy in Syria and the development of repertoires (Weeden 2002: 723).

Hence, due to all these mediums of repression, unlike their Egyptian peers, Syrians were less practised at protesting and demonstrating in the online and offline spheres before 2011 (Stacher 2012: 13). Notions such as collective identity, past protest experience and social ties that formed the repertoires of well-functioning societies were absent in Syria. My interviews with Syrian and Egyptian activists reflect the problems that this absence of repertoires created in the online and offline spheres during the 2011 protests. As the Syrian leaders lacked the broad experience in protest fields and on the Internet of their Egyptian peers, they struggled to mitigate fear amongst the Syrian public and themselves. Afraid of being caught, protesters used ICTs for their interpersonal communication rather than as organisation tools. Yet, as Chapter 6 will demonstrate, a symbiotic relationship exists between repertoires and ICT use. As protesters acquired experience in the protest fields and mitigated the fear amongst themselves, they were prepared to risk more to protect what they had achieved, resulting in them using ICTs effectively despite the risks. The protesters' hybrid media activities across and integrating new, old and face-to-face communication thus continued to improve as they used ICTs to form collective actions, creating social ties and collective identities in the online and offline spheres.

LEADERSHIP IN DIGITALLY SUPPORTED MOVEMENTS

The Catalan Sociologist Manual Castells (2010) described the recent digitally supported protests as flexible and decentralised protests, spread via the "networks" formed on the Internet. According to Castells (2010), this transformation mostly occurs as a result of trans-border network communication. Communication using ICTs helps to expand society's bonds by contributing to the formation of weak ties with strangers. Like Granovetter (1973), who claims that weak ties can in fact lead to strong bridges, Castells (2010) argues that loosely affiliated people can actually sustain and build strength over time utilising a mixture of online media and offline activities. Thus, ICTs make it easier for a movement to reach thousands or millions of people in the absence of fixed leadership and facilitate the appearance of large movements with loosely affiliated networks.

This idea of leaderless, horizontal, networked movements became a standard reference point for many communication scholars who defined the recent protests. Bimber, for instance, claims that "[a] loosely coupled network without central financing or a fixed structure for leadership, decision making and recruitment" can easily change public policy or political outcomes (Bimber et al. 2005: 370). This thesis was further developed by Bennett and Segerberg in 2013. The scholars argue that the collective action, the irreducible act that can be found at the base of traditional social movements, is changed and we need to redefine collective action in the light of technological developments. They (2013: 24) define three types of actions in the digital era. The first type is organisationally brokered collective action, which refers to conventional actions with large-scale action networks. The brokering organisation networks carry the burden of facilitating co-operation and bridging differences (Bennett and Segerberg 2013: 46). The second is a new type of action: organisationally enabled connective action. The organisationally enabled networks centre around resource-rich NGOs, but constituent organisations aim to personalise the engagement of the public. To this end, they deploy interactive media that enable the public to choose how they engage with the movement. Finally, the third is "crowd-enabled connective action" that they associate with the Arab uprisings and Occupy movements. Crowd-enabled connective actions appear due to the changing media environment and rely on ICTs. They are organised by "crowds largely without central or lead organisational actors" (Bennett and Segerberg 2013: 46). Their central organisation is conducted by spreading personal action frames on the Internet. Personal action frames refer to engagement through simple, everyday discourses shared with social networks, such as rallying cries of "Real democracy now" (Bennett and Segerberg 2013: 2).

This idea of leaderless movements has already been challenged by scholars such as Papacharissi and Oliveira (2012), Gerbaudo (2012) and Poell et al. (2015). For Papacharissi and Oliveira (2012), digitally supported movements are not leaderless. One could easily notice opinion leaders on platforms such as Twitter who direct the conversation and help to frame the events. Studying the 2011 Egyptian protests in his book, *Tweets and the Streets*, Gerbaudo aims to define these leaders and their roles in today's movements. According to Gerbaudo, the coherence of the Egyptian protests was secured not only by the micro-operations on the Internet, but also as a result of the identity, collective solidarity and soft leadership within contemporary social movements (Gerbaudo 2014: 265). Hence, the "liquid" organisation through the Internet was not totally leaderless; it was communicated by social media administrators acting as soft leaders or choreographers (Gerbaudo 2012: 145). More recently, Poell et al. (2015) define these leaders on the Internet as "connective leaders". Gerbaudo and Poell et al.'s studies showed that the presence of a common protest identity, protest culture and form of collective leadership were still vital for coherence in protest communication.

Comparing the protests in Syria and Egypt helps us to test these contradicting theories of communication scholars on the role of collective identity and leaders in the digital age. Such an examination shows the continuing importance of leaders in the organisation of today's digitally supported protests. I argue that in countries like Egypt where a vibrant civil society and past performances are present, protests are not merely formed by personalised conversations on digital media, but also by networks of leaders. These leaders differ from leaders defined by resource mobilisation theorists in the 1970s—they do not impart the image of speaking for a potential constituency (McCarthy and Zald 1977: 20). Rather, they prefer to operate invisibly, but effectively enough to give collective action a degree of coherence and sense of direction (Gerbaudo 2012: 157). In this book, the first type of leadership is classified as *soft leaders*. As defined by Gerbaudo, they act as choreographers of the protests and construct an emotional tension amongst Internet users. These leaders acquire experience in administrating a social media page, using their knowledge to set the scene for collective action. This book also demonstrates two other kinds of leadership in today's movements, namely *experienced activists* and *hybrid leaders*. While soft leaders encourage middle-class Internet users to join in the protests via their Internet activities, experienced activists operate on the ground to mobilise parts of society who do not have Internet access. Hence, unlike the professional social movement organisations of the 1970s that depended on institutional resources from private foundations, churches or business groups and had very small or non-existent membership, the experienced activists of digitally supported protests follow the tactics of classical social movement organisations. Their actions are based on indigenous leadership, volunteer staff and mass participation (Jenkins 1983).

Modern conditions have also brought hybrid leaders to the fore in the organisation of digitally supported protests. The term hybridity is used by Andrew Chadwick to define the hybrid media system of our era. This new media system is built amongst older and newer media logics. Chadwick defines these logics as "technologies, genres, norms, behaviours and organisational forms" (Chadwick 2013: 4). Hybrid leaders combine online and offline communication and organisation by harnessing newer and older media logics and trust relations in face-to-face communications to build and lead a social movement towards a strategic goal. Hence, these leaders display operational flexibility and become active in *both* online and offline spheres. While trying to mobilise society in both spheres, they also often operate as communicators or intermediaries, linking soft leaders to experienced activists. Like leaders in professional social movement organisations, they are required to know how to manipulate relevant images and support through mainstream media. They also need to be familiar with the tactical communication on the ground, which motivates the different segments of society to protest "shoulder to shoulder" (Gerbaudo 2012: 160).

This book also reveals the difficulty of finding hybrid leaders in countries where there is a lack of performances. As the public do not acquire administrative experience on social media pages, they do not develop networks in the online spheres. It is thereby challenging to find actors that can set the scene in both spheres. The co-operation of the leaders also depends on their past protest experiences. The chapters on Syria (Chapters 5 and 6) demonstrate that in countries where the social networks between different groups are not developed, as in Syria, it is difficult to find or create synergy between the mobilisation attempts of the soft leaders and the experienced activists on the ground.

Resource-Rich and Resource-Poor Activism

Finally, the last two empirical chapters of this book explore the factors that help digitally supported movements to sustain over time. As I compare a recent uprising with a negative outcome (the 2011 Syrian uprising) with another with a relatively positive outcome (the 2011 Egyptian uprising), I shed light on the sufficient elements of a successful uprising in the digital age.

What is often taken for granted in accounts of digitally supported movements is that they show robust organisational capacities in response to short-term events. For instance, Bennett and Segerberg (2013), who analyse the Occupy protests, argue that despite the absence of leaders or social movement organisations, the diversity and flexibility of these movements helped them to easily adapt themselves to unanticipated settings and rapidly developing opportunities. These movements might not always achieve impressive results, but this is due to the political context in which they operate rather than their organisational capacities (Bennett and Segerberg 2013: 193). In these types of arguments, the existence of social movement organisations and leaders is not seen as important factor for today's protests; rather, the arguments ascribe importance to suitable political context and a unifying message, which will bring about the engagement of demonstrators.

In contrast, comparing the Egyptian and Syrian uprisings will show that although the organisation of the digitally supported protests shows large differences from 1970s professional social movement organisations, their sustainability still depends on the capabilities of leaders as well as the presence of resources and political opportunities in the country. In this book, I respond to Eltantawy and Wiest's (2011) call for communication scholars to consider resource mobilisation theory in their analyses. Resource mobilisation theorists cite various internal and external resources that shape the success of the movements. Theorists such as Tilly (2006) and Tarrow (2011) put emphasis on the external resources, such as the opening and closing of political opportunities and constraints in a country. They state that the ability of protesters to use these resources depends on performances protesters learned from their past protest experiences. On the other hand, McCarthy and Zald (1977) define these resources as legitimacy, money, facilities provided by outside organisations, social networks and labour. In the digital age, ICTs have become important resources for the organisation of protests. The development of technology has reduced the need for money, labour and costs to be spent for movement organisation. ICTs enable protesters to "receive and disseminate information; help to build and strengthen ties among activists; and increase interaction among protesters and between protesters and the rest of the world" (Eltantawy and Wiest 2011: 1218). However, I argue that the availability of the Internet is not enough to create a successful movement in authoritarian states, and so the sustainability of protest movements is also dependent on other external and internal resources present in the country. These include:

a. The multiplicity of independent centres of power within the regime (a vibrant civil society with NGOs and media outlets) providing technological and logistical support and enhancing networks of protesters.
b. Getting support from different segments of the society.
c. Division amongst the social and military elites.
d. The regime's active use of hybrid media to generate a persistent message.
e. The regime's repression level.
f. Regime structure.

Scholars such as Howard and Hussein (2013) and Goldstone (2011) also add international support to this list. According to them "international power must either step in and defend the government or must constraint the government from defending itself too ruthlessly" (Howard and Hussein 2013: 119). This book only focuses on the peaceful protests in Syria and Egypt and does not explore the period when the peaceful protests turned into a conflict drawing in the international community. Hence, it does not analyse how international support affects the sustainability of the movements.

This book also puts emphasis on the capabilities of leaders to use the resources available in the country, unlike the work of Howard and Hussein (2013). As Edwards and McCarthy (2004) state, the resources of movements might lead to differences over space, through time and across constituencies. Moreover, the engagement and capacities of individuals and organisations to use these resources are crucially important in determining a movement's success and failure (McCarthy and Zald 1977: 1216). Thereby, I will argue that the reasons behind the different organisational structures, activities, actors and political outcomes of the Egyptian and Syrian protests can be explained first by the differences in resources available to the activists and second by their engagement and capacities to use them.

Another media scholar, Zeynep Tufekci (2017), who also analyses the strength of recent digitally supported protests, emphasised that today's protests develop three complex capacities, namely: to set the narrative, to affect electoral and institutional changes and to disrupt the status quo. These capacities are not the same in every movement that uses ICTs. Comparing the Egyptian and Syrian protests in Chapter 7 enabled me to show the significance of leadership skills to the movements' capacities.

Outline of the Book

A necessary analytical problem regarding digitally supported actions is the *context* in which they appear and whether these political, socioeconomic contexts affect the success and sustainability of movements. To respond to these questions, Chapter 2 explores the differing national contexts in Syria and Egypt. The political, economic and cultural contexts, as well as the civil society and media structures in these countries, are compared. It is argued that the different political contexts had cumulative impacts on the progression and outcomes of both protest movements. Political opportunities that were not opened in Syria, such as the scepticism of the Egyptian army towards Gamal, the son of Hosni

Mubarak, and the cluster of new elites around him, differentiated the political scenes in Syria and Egypt. They also affected the progression of both protest movements.

Chapters 3 and 4 explore the Egyptian uprising. Chapter 3 explores the types of actors and their repertoires of contention in the organisational process of the uprising. Here, it tests the recent theories on leadership in digitally supported movements and develops them by introducing three types of leadership found in the Egyptian protests. Moreover, the chapter shows that the "past protest experience" of Egyptians and their "social networks" were crucial for coherent organisation. Unlike in Syria, the actors worked in synergy in the organisational process throughout the Egyptian uprising.

Chapter 4 analyses in depth the validity of "resource mobilisation theory" in today's movements. While it traces the mobilisation, co-ordination and sustainment attempts of the Egyptian protesters, it also pays attention to the role of the political institutions and NGOs in enlarging the networks and providing technical resources. The first finding of this chapter shows that the past protest experience of the activists enabled them to effectively use interpersonal communication methods in both the online and offline spheres. Based on their past protest skills, protesters utilised face-to-face communication to mobilise the poor segment of the society. ICTs were used to mobilise middle-class Internet users, buffer police officers, co-ordinate different groups and internationalise the movement. The findings of this research build on the current literature by showing that the effectiveness of ICTs depends not only on the past experience of the actors involved but also on the availability of organisations that provide facilities for the movement. The civil society in Egypt, along with political institutions, NGOs and human rights organisations, shaped the way in which ICTs were used in the Egyptian protests. Thanks to this organisational support, Egyptians also managed to create a non-sovereign sphere in Tahrir in which different ideological groups could coexist, pray and protest together. Hence, both Chapters 3 and 4 prove the validity of resource mobilisation theory in today's movements.

Chapters 5 and 6 focus on the Syrian uprising: like the method applied in the Egyptian case, Chapter 5 starts by analysing the types of actors and their repertoire of contention in the organisation process. It questions why the "soft leaders" could not become successful in creating the scene for the protests in Syria, even while experienced activists did so on the ground. Which mobilisation factors become apparent in the Syrian uprisings and what were their roles in the movement? Analysis shows that when the society does not acquire enough experience in the protest field and reinforced repertoires, administrators cannot mobilise the public via the loose networks formed on social media. "Grievance", "repression" and "social networks" become key mobilising factors that activate the public. Following the organisation process over time, Chapter 6 explores the mobilisation, co-ordination and sustainment attempts of the Syrian protesters and their repertoires of contention. The chapter shows that there is a symbiotic relationship between the repertoires and the ICTs. It also traces the interaction between the state and dissidents and identifies the activities of the state in the online/offline spheres, as well as the mass media.

Chapter 7 compares the capabilities of the leaders and the organisational structure of these in Egypt and Syria. I also question whether these fit into any of the three models that Bennett and Segerberg (2013) define in their book, *The Logic of Connective Action*, namely crowd-enabled connective action, organisationally enabled connective action and organisationally brokered connective action. The findings presented in this chapter show that the Syrian and Egyptian protests essentially reflected different organisational structures. While there was a coherent and centralised organisation in the Egyptian protest, with a strong repertoire, the Syrian protest reflected a decentralised structure with a weak repertoire. Moreover, the Egyptian protests were not personalised and loosely co-ordinated actions, corresponding to the crowd-enabled movement definition of Bennett and Segerberg (2013). They reflected the features of organisationally enabled movements, an action type that lies between the organisationally brokered and crowd-enabled types (Bennett and Segerberg 2013: 48). On the other hand, the Syrian protests better represent the features of a crowd-enabled movement, with no formal organisation or co-ordination of action. Yet, unlike Bennett and Segerberg's (2013) definition of crowd-enabled movements, technology was not the prominent agent of mobilisation in the Syrian uprising.

Finally, the common goal of both Egyptian and Syrian movements was to mount a revolution, a rapid and systemic political change through a social movement (Cole 2014: 267). By referring to the views of my interviewees and the secondary data collected for this research, I will discuss what the sufficient factors to achieve this goal were.

The concluding chapter (Chapter 8) primarily interrogates contemporary arguments on the role of leaders, organisations and ICTs in recent digitally supported movements to provide an overview of the changing conception of the collective action. It then retraces the central arguments and findings of the book outlined in each chapter and their contributions to the field. Further on, I point to the need to explore these arguments within other contemporary protests to develop our understanding of digitally supported movements and the factors that lead to their long-term adaptation, change or decline. Finally, by referring to the arguments of my interviewees, I discuss the challenges and problems that digitally supported movements face today and their contributions to social and political change in non-democratic countries.

Notes

1. Interview with Nisrin (pseudonym), Egyptian activist, London, 2013.
2. Interview with Mona (pseudonym), Egyptian activist, Cairo, 2014.
3. "The 2011 Egyptian protests" were used in this book in reference to the protests in Egypt between 25 January and 11 February 2011.

References

Abaza, J., & Youssef, A. (2011, January 28). *28 January 2011: Egypt's Bastille Day.* Retrieved from https://bit.ly/2KXG4MV.

Alaoui, H. (2016, April 12). *Is the Arab world better off, five years after the Arab Spring?* [Web log post]. Retrieved from http://macmillan.yale.edu/news/arab-world-better-five-years-after-arab-spring.

Alvarez-Ossorio, I. (2012). Syria's struggling civil society. *Middle East Quarterly, 19*(2), 23–32.

Anckar, C. (2008). On the applicability of the most similar systems design and the most different systems design in comparative research. *International Journal of Social Research Methodology, 11*(5), 389–401.

Aouragh, M., & Alexander, A. (2011). The Arab Spring: The Egyptian experience: Sense and nonsense of the Internet revolution. *International Journal of Communication, 5,* 1344–1358.

Bennett, W. L., & Segerberg, A. (2013). *The logic of connective action.* Cambridge, UK: Cambridge University Press.

Bimber, B., Flanagin, A., & Stohl, C. (2005, November). Reconceptualizing collective action in the contemporary media environment. *Communication Theory, 15*(4). https://doi.org/10.1111/j.1468-2885.2005.tb00340.x.

Castells, M. (2010). *The rise of the network society*. Oxford, UK: Blackwell.

Castells, M. (2012). *Networks of outrage and hope: Social movements in the Internet age*. Cambridge, UK: Polity Press.

Chadwick, A. (2013). *The hybrid media system: Politics and power*. New York, NY: Oxford University Press.

Cole, J. (2014). *The new Arabs: How the millennial generation is changing the Middle East*. New York, NY: Simon & Schuster.

Edwards, B., & McCarthy, J. (2004). Resources and social movement mobilization. In D. A. Snow, S. A. Soule, & H. Kriesi (Eds.), *The Blackwell companion to social movements* (pp. 116–153). Malden, MA: Blackwell Publishing.

Eltantawy, N., & Wiest, J. (2011). Social media in the Egyptian revolution: Reconsidering resource mobilization theory. *International Journal of Communication, 5*, 1207–1224.

Faris, D. (2013). *Dissent and revolution in the digital age*. New York, NY: Palgrave Macmillan.

Gerbaudo, P. (2012). *Tweets and the streets: Social media and contemporary activism*. New York, NY: Pluto Press.

Gerbaudo, P. (2014). The persistence of collectivity in digital protest. *Information, Communication & Society, 17*(2), 264–268. https://doi.org/10.1080/1369118X.2013.868504.

Gladwell, M. (2010, October 4). Small change: Why the revolution will not be tweeted. *The New Yorker*. Retrieved from http://www.newyorker.com/magazine/2010/10/04/small-change-malcolm-gladwell.

Goldstone, A. J. (2011). Understanding the revolutions of 2011. *Foreign Affairs*. Retrieved from https://www.foreignaffairs.com/articles/north-africa/2011-04-14/understanding-revolutions-2011.

Granovetter, M. (1973). The strength of weak ties. *American Journal of Sociology, 78*(6), 1360–1380.

Heaven, W. (2009, July 8). *Iran's crackdown proves that the 'Twitter revolution' has made things worse* [Web log post]. Retrieved from http://blogs.telegraph.co.uk/news/willheaven/100002576/irans-crackdown-proves-that-the-twitter-revolution-has-made-things-worse/.

Herrera, L. (2014). *Revolution in the age of social media*. London, UK: Verso.

Howard, P., & Hussain, M. (2013). *Democracy's fourth wave? Digital media and the Arab Spring*. New York, NY: Oxford University Press.

Howard, P., & Parks, M. (2012). Social media and political change: Capacity, constraint, and consequence. *Journal of Communication, 62*(2), 359–362. https://doi.org/10.1111/j.1460-2466.2012.01626.x.

Jenkins, J. C. (1983). Resource mobilisation theory and the study of social movements. *Annual Review of Sociology, 9*, 527–553.

Leverett, F. (2005). *Inheriting Syria: Bashar's trial by fire*. Washington, DC: Brookings Institution Press.

Lijphart, A. (1975). The comparable cases strategy in comparative research. *Comparative Political Studies, 8*(2), 158–177.

Lim, M. (2012). Clicks, cabs, and coffee houses: Social media and oppositional movements in Egypt, 2004–2011. *Journal of Communication, 62*(2), 231–248. https://doi.org/10.1111/j.1460-2466.2012.01628.x.

McCarthy, J. D., & Zald, M. N. (1977). Resource mobilization and social movements: A partial theory. *American Journal of Sociology, 82*(6), 1212–1241.

Morozov, E. (2012). *The net delusion: How not to liberate the world.* London, UK: Penguin.

Papacharissi, Z., & Oliveira, M. (2012). Affective news and networked publics: The rhythms of news storytelling in Egypt. *Journal of Communication, 62*(2), 266–282. https://doi.org/10.1111/j.1460-2466.2012.01630.x.

Phillips, C. (2015). Sectarianism and conflict in Syria. *Third World Quarterly, 36*(2): 357–376. https://doi.org/10.1080/01436597.2015.1015788.

Poell, T., Abdulla, R., Rider, B., Woltering, R., & Liesbeth, Z. (2015). Protest leadership in the age of social media. *Information, Communication and Society, 19*(7), 994–1014.

Rihoux, B. (2006). Qualitative comparative analysis and related systematic comparative methods. *International Sociology, 21*(5), 679–706. https://doi.org/10.1177/0268580906067836.

Rosen, J. (2011, February 13). The Twitter can't topple dictators. *Press Think.* Retrieved from http://pressthink.org/2011/02/the-twitter-cant-topple-dictators-article/.

Seawright, J., & Gerring, J. (2008). Case selection techniques in case study research: A menu of qualitative and quantitative options. *Political Research Quarterly, 61*(2), 294–308.

Shirky, C. (2009, December 11). The net advantage. *Prospect Magazine.* Retrieved from http://www.prospectmagazine.co.uk/magazine/the-net-advantage.

Stacher, J. (2012). *Adaptable autocrats: Regime power in Egypt and Syria.* Stanford, CA: Stanford University Press.

Tarrow, S. (2011). *Power in movement: Social movements and contentious politics.* New York, NY: Cambridge University Press.

Tilly, C. (2006). *Regimes and repertoires.* Chicago, IL: University of Chicago Press.

Tufekci, Z. (2017). *Twitter and tear gas: The power and fragility of networked protest.* New Haven, CT: Yale University Press.

Weeden, L. (2002). Conceptualizing culture: Possibilities for political science. *The American Political Science Review, 96*(4): 713–728.

CHAPTER 2

Egypt and Syria: Similarities and Differences Between Two Countries

For many years, both the Syrian and Egyptian regimes proved themselves to be remarkably enduring. In 2007, for instance, upon Bashar al-Assad's election for another seven-year term, Eyer Zisser, author of *Asad's Legacy, Syria in Transition*, claimed that the Syrian regime was more stable than ever. In late 2010 and 2011, Syria still had this reputation and seemed to be a fairly stable state (Lesch 2012: 38). Like the Syrian regime, the Mubarak regime in Egypt was widely viewed as out of touch yet stable. Power had been monopolised around the president for thirty years through rigged elections. Moreover, the regime had long been successful in managing participation in formal political life by applying patronage, elaborate legal restrictions and coercive power to divide and punish opposition forces (Al Aswany 2011: viii). Therefore, it is important to ask: What are the factors that undermined the stability of both regimes and led to the outbreak of widespread protests? In this chapter, I compare the political, social, economic and communication structures in Syria and Egypt before 2011 and discuss the factors that led to widespread protests in both countries.

The chapter shows that both regimes shared similar historical paths; they both advanced populism and rhetorically advocated Arab nationalism in their foreign policies. Earlier, Nasser had intensified Egyptian populism by increasing the role of the state in the economy in Egypt.

B. Aslan Ozgul, *Leading Protests in the Digital Age*,
Palgrave Studies in Young People and Politics,
https://doi.org/10.1007/978-3-030-25450-6_2

Similarly, Syria's Ba'ath Party leaders pursued populist policies when they came to power. Moreover, both countries had military officials in leadership positions between 1963 and 1970. They continued to resemble one another during the Anwar el Sadat and Hafiz Al-Assad periods as well (Stacher 2012: 163).

Yet, this chapter reveals that despite their similarities, one could also observe structural differences in these two authoritarian states. To understand the reasons behind the 2011 Syrian and Egyptian uprisings, it is crucial to analyse the nuanced political, social and economic factors that differentiated each country. Thus far, few studies have compared the different structural factors in both countries. Stacher's book, Adaptable Autocrats (2012), is one of the few studies that conducted a broad analysis of the political structures and the operation of power in Egypt and Syria. In his book, Stacher argues that "Egyptian elites maintained centralised authority while Syrian power was decentralised heading into and emerging out of their uprisings" (Stacher 2012: 4). According to Stacher (2012), the centralisation of power in Egypt facilitated the swift transition of the ruling coalition. This chapter also addresses the political structure in the two countries. Additionally, it focuses on the repertoires of contention, media and civil society of Egypt and Syria, as I argue that these factors were also significant in the differing outcomes of protests. Arab society is often deemed as having pressure groups that had pressed for greater democracy before 2011. Comparing the civil society in Syria and Egypt will enable us to see that, faced with fierce repression, the intellectuals in Syria could not find enough of a chance to operate in the political sphere as their peers in Egypt. As a result, they lacked the political experience and repertoires necessary to enact change.

On the other hand, unlike the Syrians, the Egyptians had necessary political opportunities for an uprising and a suitable political environment for a revolution. For instance, the wall of fear had already been broken down for 10 years and this helped the Egyptian public to create virtual networks that unified against the Mubarak regime before the protests. Furthermore, the networks formed by past collective actions of middle- and working-class segments assisted the appearance of well-connected experienced activists on the ground. Unlike Syria, the developed civil society and multiparty system of Egypt also helped the emergence of new political opportunities for collective action in Egypt.

POLITICAL STRUCTURE

Syria's Decentralised Political Structure

The political structure of Syria can only be understood through analysis of the founding ideology behind it, classic Ba'athism. Essentially, classic Ba'athism adopted a unifying role. The founders of the Ba'ath Party aimed to awake the slumbering Arab nation and to unify its populace, which had been divided by imperialism into many states. In this respect, when the Ba'ath Party was founded, its founders initially intended to bridge the class and sectarian cleavages that divided Syrians and had kept them weak for decades (Hinnebusch 2001: 31). Prior to the early 1960s, Ba'ath Party leaders did also not favour one community over the others. The Syrian Alawi community had a relatively humble status as a minority community, just as Shiis, Druzes, Ismailis, Greek Orthodox, Maronites and Christians also had in Syria (Haddad 2012). The Alawis were at the bottom of the social ladder and in the margins of Syria's political and economic system.

This political structure changed with the neo-Ba'ath Revolution in February 1966, which transformed the Alawi community into one of the major concentrations of power in Syria. Following the Corrective Revolution of November 1970, Hafez Al-Assad took the presidency and subsequently empowered the Alawi officer corps as he appointed fellow Alawis to key leadership positions (Zisser 2001: 19–20). Thus, unlike previous occasions, in which opposing forces had competed for power, when Hafez Al-Assad took power in 1970, the Syrian political elite showed a high degree of sectarian and regional homogeneity (Dam 1996: 137). Nikolaos Van Dam (1996), who examined these sectarian regional and tribal loyalties in Syria, claimed that at first, Hafez Al-Assad sought to achieve a secular, Arab nationalist Syria similar to the vision of the founders of the Ba'ath party, but in practice, he could not successfully follow this ideal due to the realities of Syrian power politics. For practical purposes, Ba'athist leaders depended on people from their own communities and these people shared little of their ideology. Thus, once they attained power, the Alawis, Druzes or whichever other minority it happened to be favoured members of their own communities (Dam 1996: 138–140).

However, it is essential to note that the roots of the Assad regime did not only depend on the Alawi corps. Hafez Al-Assad also created a military mercantile complex through strong ties between the government and the large Sunni business class in the country. This support of Sunni elites had major influence on the stability of the regime. Thanks to this close relation, the regime could even suppress one of the biggest threats to its existence: the Muslim Brotherhood uprising, which had begun in the 1970s and had resulted in 10,000–20,000 deaths.

Moreover, the solid base of the regime was also the result of the social and political revolution which took place in Syria after the Ba'ath party seized power in 1963 (Zisser 2001: 20). Raymond Hinnebusch (2001) classifies this regime type as populist authoritarian, which embodies a decolonising, state-building strategy. This type of regime attacked the old dominant classes and sought legitimacy through egalitarian ideology and the political integration of the middle and lower strata. It was populist in the sense that it challenged the status quo, rather than defending the traditional. Regarding the structure, the army had been a central pillar of the regime, but reflected an "army-party symbiosis" rather than singular military rule (Hinnebusch 2001: 1).

When he came to power in 2000, Bashar al-Assad inherited this decentralised political field, where elites from more than one institution, or repository of power, had an impact on coalition change, policy implementation and cohesion (Stacher 2012: 81). As with his father, Bashar al-Assad has enjoyed the highest decision-making authority within the regime. He was already appointed as commander-in-chief of the Syrian army and general secretary of the Ba'ath party before attaining the presidency. As president, he has led the Progressive National Front, a formal alliance that brings a number of smaller, tolerated parties together with the leading Ba'ath Party. Realising the need to pre-empt calls for democracy, Hafiz Al-Assad introduced in 1990 a new law by which voters could actually choose between the list of the National Progressive Front and individual independent candidates during elections. However, every candidate needed to be approved by the authorities, which gave the president additional power over the opposition (Perthes 1992). Bashar al-Assad had also the power to appoint and dismiss the cabinet, judges, provincial governors, university presidents and other senior officials. Moreover, he could dissolve parliament, accept or deny parliamentary decisions and has legislative powers of his own (Perthes 2004: 7).

Nevertheless, when he comes to power researchers such as Zisser (2001: 15) criticised Bashar for not showing charisma and leadership qualities and argue that Bashar al-Assad lacked maturity, experience and self-confidence of a leader. In addition, due to the decentralised structure of the political system Bashar had to share all his power with the old guard entrenched in the Ba'ath party and who were wary of his project. This in turn increased the question of whether Bashar al-Assad ruled Syria or the elites in the Ba'ath party (Zisser 2001: 15).

To better comprehend the political structure, we must also understand the different types of elites in the Ba'ath party. Volker Perthes, author of *Syria Under Bashar al-Assad* (2004: 14), defines three types of Syrian regime elites in the Bashar al-Assad period: the first is the conservatives who desired to maintain the status quo in Syria; the second is the democratic reformists who oppose the conservatives and are critical of the entire system and demand fundamental change. Finally, the third type, the modernisers, amongst whom Assad belongs, sought mere technological modernisation and gradualism. Unlike the democratic reformists, this group did not seek to reform the entire system. Instead, they created a new kind of modernised authoritarianism in Syria (Perthes 2004: 6).

In order to exercise his authority over elites, Bashar al-Assad entered an extended struggle with the established elite and used his power to force the old generation into retirement. In their place, he inserted his loyalists in the army and security forces and placed reforming technocrats into the government (Hinnebusch 2012: 99). The reforming technocrats were mostly chosen by members of the Syrian Computer Society, of which Bashar al-Assad was the head from 1995 to 1999. Thanks to this renewed political and administrative elite, Assad formed his own network of loyalists and it was seen that, in this manner, his personal choices and priorities were best represented in the cabinet (Lesch 2012: 2).

Egypt's Centralised Executive Authority

Unlike Syria, the institutions in Egypt operated within a centralised structure. While the security services had an apolitical character, the ruling NDP was weak. The executive branch had a centralised authority over state organisations and dominated the political order (Stacher 2012: 38). In addition, Mubarak inherited a system where there were no obvious figures visible in military and political circles who could either block or challenge his formal appointments (McDermott 1988: 73). Hence, Sadat's assassination created a favourable environment for Mubarak and gave him the opportunity to heal the country's wounds

(Cook 2012: 158). While he restored calm, he repressed any potential challenge from extremist groups, such as Muslim fundamentalists and fanatics (McDermott 1988: 73).

Despite this unchallengeable executive power, in the early 1980s Mubarak had difficulty consolidating his authority. According to the journalist Anthony McDermott (1988), Mubarak's weakness was linked to his position as a caretaker president. Previous presidents had achieved significant accomplishments. For example, Nasser came to power via a coup and legitimised his power via his achievements on behalf of Egypt. When Nasser died, Sadat established his legitimacy by heading off the Left and claiming to put Egypt's interests first by signing a contentious peace treaty with Israel and ending the 1973 war. Unlike the two former leaders, Mubarak inherited power from Sadat rather than seizing it and attached importance to continuity. He ensured the stability of the system and acquired his legitimacy "through the unexciting policy of buying time" (McDermott 1988: 74–75).

On the other hand, Steven A. Cook (2012), an expert on Arab and Turkish politics, argues that the problem arose from Mubarak himself. In Egyptian society, there was a sense Mubarak was weak. The defence minister, Abdel Halim Abu Ghazala, who combined ambition and popularity within the ranks of the armed forces, was seen as a rival for Mubarak's position. In 1989, Mubarak would solve this problem by removing Ghazala from power, replacing him with Field Marshal Mohamed Hussein Tantawi in 1991. As Tantawi was not a threatening figure, he would stay in office until the end of Mubarak's regime (Nassif 2013: 515).

The power in the parliament is also vested in the president. Egypt enjoyed a multiparty system with relatively free elections since the 1970s (Kramer 1992: 23). Despite his unexciting policies, the ruling National Democratic Party (NDP) controlled at least three-quarters of the seats during Mubarak's reign. In 1995, this number rose to 94 per cent, and in 2010, 97 per cent of the seats were secured by NDP members. The election system had an important role in this success. Local councils were elected in accordance with a winner takes all system. In return, this guaranteed the NDP's monopoly of power (Lesch 2011: 35).

The regime also sought to gain the support of the public by making good on the NDP's commitment to reform (Cook 2012: 173). Under promises of modernisation, Mubarak initiated important political reforms in 2005 and 2007. In 2005, Article 76 of the constitution was changed. As a result, plebiscite in the National Assembly was replaced by direct elections and multiparty presidential elections were allowed. In 2007, further changes took place with the amendment of 34 different articles in the constitution. These changes led to an alteration in the rhetoric of many constitutional provisions from socialism to a more market-orientated philosophy. Moreover, while the power of the executive was restrained, more power was granted to the parliament. As part of this power-shift, the People's Assembly was charged with approving the annual budget (Sika 2012: 183).

These reforms always come with other restrictions that have empowered executives. For instance, in order to be eligible to compete for the presidency, independent candidates were required to collect "sixty-five signatures of support from the People's Assembly, twenty-five signatures of support from the Consultative Council and ten members of municipal councils in at least fourteen of Egypt's twenty-seven governorate" (Cook 2012: 185). Candidates from legal parties are challenged with a different set of constraints. Presidential candidates, for instance, can only be nominated by a party that has been licensed for the previous five years and held five per cent of the seats in the House and Parliament (Cook 2012: 185). Moreover, only a person who had held a senior leadership position in the party for at least a year before the elections could run for office (Sika 2012: 184).

Along with these amendments, the regime extended the state of emergency, which reinforced the president's absolute authority, by empowering him and allowing him to dissolve the parliament at any time without having to answer to the public through referenda (Cook 2012: 187). New laws were also promulgated to exclude opposition parties from local councils and tightened the control of the ruling NDP over parliament (Lesch 2011: 35–36). The regime also allocated specific powers to the minister and Minister of Interior and allowed them to restrain "the movement of individuals, search persons or places without warrants, tap telephones, monitor and ban publications and intern suspects without trial" (Lesch 2011: 36). New restrictions also concerned groups that aimed to form a political party. For instance, if a group wanted to be considered a political party, they were required to represent a new addition to the political arena. Hence, if there was already a party on the left, the emergence of another leftist party was not allowed in the Egyptian political system (Cook 2012: 186). With the revisions, the political activities of the religious groups and parties based on religion were also forbidden. This led to the ban on the Muslim Brotherhood from all political activity (Cook 2012: 188).

Mubarak further secured his unchallengeable position in the system by being able to remove or move any person, replacing any who did not submit to his rule. In doing so, he eroded any organisational, personal or institutional resilience that surfaced (Stacher 2012: 83). When it came to the army, Mubarak always upheld the privileges of the military elite. Amongst the seven commanders of the Republican Guard who served under Mubarak, four were appointed as governors. Sustaining the interests of the armed forces' leaders was key to sustaining civil–military relations (Nassif 2013: 509). In pursuit of this aim, a portion of the public budget was distributed amongst military and police leaders in direct cash instalments. From the highest echelons of the armed forces, this money, which was referred to as loyalty bonuses, trickled down to senior officers who held the ranks of major general and brigadier general (Nassif 2013: 527). Thus, Mubarak's regime made the military elites wealthy and unanswerable to the law in return for their loyalty. By being in command of the army and parliament, Mubarak could easily check the development of autonomous elites and institutions (Stacher 2012: 83). This in turn gave him centralised executive power by which he could eradicate any potential challenges.

Economic Structure

The Modernised Authoritarianism of Bashar Al-Assad

Since independence, the state was positioned as the dominant economic actor in Syria and pursued a form of statist economic development. During the Hafiz Al-Assad period, this statist model assigned priority to the corporatist logic of inclusion, stability and state dependence rather than economic development (Abboud 2015). The statist model eventually exhausted itself, as it promoted consumption at the expense of accumulation (Hinnebusch 2009; Abboud 2015). Consequently, the state engaged in a series of economic reform in the late 1980s that continued with acceleration in 1990s.

At first glance, the economic structure of Syria was in a good condition during the Hafez Al-Assad period. Many academics and journalists praised his stable regime and regarded him as a successful and talented leader. The first proponent of this argument was Patrick Seale (1998) who argues that Hafez Al-Assad's leadership transformed the Syrian state and turned it into a regional power. Similarly, Hinnebusch (2001) describes Hafez Al-Assad as a determined, intelligent leader and dedicated to his mission. Indeed, over the course of Hafez Al-Assad's presidency, Syria was faced with serious challenges. These included the Islamic uprising of the late 1970s, the economic stagnation of the 1980s, which brought an end to the Cold War era's Soviet aid and protection, and waves of economic globalisation and democratisation, which posed additional challenges for the regime during the same time period. Despite these challenges, the regime succeeded in retaining remarkable stability in its economic and foreign policies (Hinnebusch 2001: 1). Yet, although Assad succeeded in stabilising the country after a period of unsteadiness, his intelligence and dedication to his mission could not prevent Syria from experiencing deep social and political problems. Zisser (2001: xiii), for example, who questioned the success of Hafez Al-Assad, asked the question: "to what extent did Assad's image as a talented, successful and indeed omnipotent leader correspond with reality?"

Contrary to other accounts, Zisser (2001) argues that when he died in June 2000, Hafez Al-Assad left his son, Bashar al-Assad, a Syria that was in the throes of deep social and economic distress and the country was isolated, both internationally and regionally (Zisser 2001: xiii). According to Zisser (2001), the reason behind this failure of the regime was the lack of modern reforms it enacted in the country. The limited measures of Hafez Al-Assad's leadership brought no real and encompassing answer to Syria's problems (Zisser 2001: xv).

Another problem that Bashar inherited from his father was the high level of corruption in the governmental system. During the presidency of Hafez Al-Assad, corruption was rife in both the party apparatus and regime. The political and military elite largely used their power to enrich themselves. Thus, when Bashar al-Assad assumed power in 2000, he received from his father a country in a dilapidated condition, characterised by a stagnant economy, pervasive corruption and political repression (Lesch 2012: 5). Due to all of these problems, there was a growing desire for reforms in the public. When Bashar al-Assad assumed the presidency, in response to these desires, he emphasised his determination to modernise the country. He presented himself to the public as a reformer, a fighter of corruption and advocate of modern technology (Magout

2012). The young leader became a hope for Syrians and was even called "the hope" by the public (Lesch 2012: 2).

During the Bashar al-Assad period, the state's move towards a market economy accelerated through gradual, selective and strategic economic liberalisation (Abboud 2015). As part of the reforms, the government shifted public sector responsibilities to the private sector. For instance, all new public sector employment ceased; this act removed a major source of social mobility and economic stability for 25–30 per cent of all Syrians who were employed in the public sector. The declining oil revenues also forced the government to reduce the state's budgetary commitments, such as public spending and investment. With regard to social spending, the government adopted a tenth five-year plan for fiscal discipline. The plan called for the abolition of expenditures that had distorted the prices and led to inefficiencies in the production cycle. It also called for the reduction and restructuring of public sector enterprises. The government had embarked on the privatisation of universities and colleges. It also removed subsidies on agricultural inputs, especially fuel, oil, seeds and fertilisers. This led to the decline of agriculture's contribution to the national economy. Finally, it introduced price liberalisation for products including those essential to everyday life, such as heating oil or foods as potatoes. For most low-income families, these products became unaffordable.

Hence, the succession from Hafez to Bashar al-Assad did not live up to expectations. Bashar al-Assad and his administration adapted an agenda that was not fundamentally opposed to that of their predecessors. Despite the gradual process of reform, the state failed to provide a comprehensive market economy model. What was seen, rather, was a renewal of the interventionist role of the state, which repositions itself as the guarantor of social stability and welfare. Abboud (2015) argues that there was recognition of the necessity of the market economy model, but how the government could cede economic authority to the market while still protecting social stability and welfare was ambiguous.

There might be two reasons behind the continuity of the same agenda in Syria. One is explained by Stacher (2012), who argues that the diffusion of power amongst the branches of government reduced the decision-making capabilities of Bashar al-Assad and hindered the modernisation process. The countervailing domestic interests of a small group of oligarchs thus constrained attempts at reform. On the other hand, Perthes (2004) reminds us that Assad was a moderniser rather than a reformist, so instead of introducing reforms he established a new kind of modernised authoritarianism in Syria.

The Market Economy of Hosni Mubarak

Like Bashar al-Assad, Mubarak would start his reign in the context of severe economic crisis (Soliman 2011: 38). Sadat's economic opening in 1974 gave way to a transition from Nasser's state socialism to a freer form of economy (Marsot 2007: 158). Nevertheless, the Sadat regime did little to introduce the necessary rules, regulations and laws that are crucial for successful modern economies (Cook 2012; Marsot 2007). Hence, like Hafez Al-Assad, Sadat left Egypt on the edge of a severe crisis. By looking to Egypt's national income, which was slightly less than 900 US dollars per capita in the late 1980s, the gravity of the economic conditions was apparent (Cook 2012: 159; Soliman 2011: 36). Mubarak was quick to identify the economy's long-standing problems, such as the population growth rate, but was slow to take action to manage this. In the meantime, the new economic climate led to the emergence of thousands of millionaires, widened the gap between rich and poor, promoting class divisions and tensions (McDermott 1988: 82). Critically, in the mid-1980s to late 1980s, oil prices and revenues started to decline, which produced a massive crisis for an Egypt that depended largely on oil revenues and labour remittances (Pfeifer 2012: 205).

In the intervening period, the economic leadership of Mubarak was mostly described as inefficient in reversing the sharp drops in revenues. Theorists such as McDermott (1988: 78–82) criticised the regime for not reforming the country's extremely inelastic tax system by which the rich saw soaring incomes, benefiting from the *infitah* (open-door policy of Sadat), with the government gaining little from it. This was only example of internal reforms, which could have been introduced by the Mubarak regime. Nevertheless, in 1989–1990, the Mubarak regime could not withstand pressure from the International Monetary Fund (IMF). Price increases were introduced for electricity, some petroleum products, fuel, flour and oil. In the 1990s, the IMF also pushed for the Mubarak government to privatise state-run businesses (Hopwood 1992: 193).

The close co-operation between the IMF and Egyptian government reflected one of the main differences between their foreign policies and, consequently, the economies of Syria and Egypt. Bashar al-Assad adopted a foreign policy congruent with public opinion and subscribed to Ba'athist Arab nationalist doctrine. Thus, he positioned his country against US and Israeli interests. Syria used its close relations with Hezbollah to pursue its own strategic goals, such as convincing Israel to negotiate peace on a basis acceptable to the Assad regime and strengthening Syria's dominant position in Lebanon. Moreover, during the Iraq war, Syria did not meet US demands, such as facilitating the movement of the US army across the border into Iraq (Leverett 2006: 17). Unlike socialist Bashar's anti-American foreign policy, Sadat turned Egypt into a strategic partner of the USA and this alliance had a subsequent impact on the Egyptian economy during the Mubarak era. For instance, throughout the 1980s and 1990s, Washington poured resources into Egyptian social and economic development (Cook 2012: 160–162). Egypt also received a record amount of aid ($4.8 billion in 1990–1991) as a reward for participating in the 1991 war to expel Iraqi forces from Kuwait (Pfeifer 2012: 206).

With the help of this financial aid, the Egyptian economy was in a far better condition by the 2000s, with income inequality and poverty measurements showing an improved situation. The chart below shows the GDP growth in Egypt per years.

Economic improvement had then arisen from typical circumstances, such as the US invasion of Iraq and subsequent rise in petroleum prices, as well as the consequent rise in Suez Canal revenues, all of which helped Egypt to get out of its financial troubles (Soliman 2011: 49). Like the economic modernisation attempts of Bashar al-Assad, the Mubarak government also began to establish a market economy, which had the potential to generate lasting growth (Cook 2012: 174). Abdelrahman (2014) describes the Egypt of the 2000s as a poster child for a new liberal order in the region, which introduced reforms such as the privatisation of public assets, the launch of drastic cuts in social expenditure, legal reforms in 2003 to guarantee flexible employment and the privatisation of agriculture. The state also removed trade barriers and generally prioritised the interests of capital (Abdelrahman 2014: 7).

Gamal Mubarak, considered the likely successor of his father, also played an important role in the establishment of this neoliberal order. He reorganised the NDP as the stronghold from which a new local version of the neoliberal order was promoted. Just like Bashar al-Assad, the young Mubarak created a policy committee in 2002, surrounding himself with a group of reformers. These reformers shared Gamal's passion and worked to scrap the remaining elements of the developmentalist state (Abdelrahman 2014: 7). Nevertheless, the improvements in the neoliberal order were not enough to attract the necessary investment to the economy (Abdelrahman 2014: 7). In response, in July 2004, the cabinet reshuffled and a businessmen cabinet emerged in the lead of Ahmed Nazif (Lesch 2011: 39). Hence, Egypt witnessed the rise of a new political class that dominated the policy process.

The resulting oligopoly actually had negative effects on prices and also grave consequences for industrial production in the country. While privatisation had been an advantage for the Egyptian business class, it increased unemployment and created an ever-increasing income gap in Egyptian society.

The neoliberal order also had an impact on the value of the Egyptian pound. Floating the Egyptian pound made it subject to global market forces; its value fell. The price of food and consumer products increased overnight and affected the daily life of the working and lower classes in 2008 (Cook 2012: 176–177). Despite these problems, Egypt's economy has generally been praised by experts for not having been much affected during the financial crisis of 2008–2009 (Pfeifer 2012: 206). However, the Egyptian Scholars, Abdelrahman (2014: 11) and Soliman (2011: 52) note that the success of the Egyptian economy derived from rentier incomes. These rentier incomes were labour remittances, Suez Canal fees and high-energy prices, especially of natural gas and petroleum, over which the regime had little control. Egypt might easily find itself on the edge of economic crisis during a crisis of rentier incomes.

Civil Society and Press

The Damascus Spring and Its Impacts

Modernisation in Syria could in particular be seen at the social level at the beginning of the 2000s. The country enjoyed progressive reforms, which gave way to more political openness in the country. Some of those reforms were general amnesties for political prisoners and the licensing of private newspapers, such as Ad-Dommari (the first privately owned newspaper since the 1963 Ba'ath coup). Moreover, a re-organisation in the state-controlled media apparatus was undertaken, with open criticism and dissent more tolerated (Lesch 2012: 8). In this reforming period, Syria's media developed, becoming more pluralistic and exacting milder censorship. Internet access increased. At the beginning of 2003, the Internet connection rate was extremely low, with only 1.45 per cent of Syrians connected. Noting this deficiency, the government introduced computer lessons in schools and announced its intention to provide the Internet to at least five per cent of the population (Perthes 2004: 33). Nine years later, Internet penetration had risen to 22.5 per cent.[1] Satellite television was introduced during the second part of the 1990s, and television imports were legalised. This allowed Syrians to watch Arabic-language satellite TV services, such as Qatar-based Al Jazeera, which became a major source of news for Syrians (Perthes 2004: 20; George 2003: 134). Moreover, the pan-Arab *al-Hayat* and *al-Sharq al-Awsat* newspapers, as well as Lebanese, Gulf Arab and Jordanian media, were regularly made accessible to the public (Perthes 2004: 20). Since 2001, the leading parties could publish their own newspapers. They provided liberal criticism through their press and demanded faster and more extensive reforms. A handful of privately owned magazines were published, such as *al-Doumari*, which was owned by the respected cartoonist, Ali Ferzat, the business-orientated *al-Iqtisadiyya*, the political news magazine *Abyad wa Aswad* and a society magazine called *Layalinan* (George 2003: 121).

On the other hand, the reform period, labelled "Damascus Spring", led to rapid growth in the number of civil society organisations and pro-democracy groups (Lesch 2012: 8). These new actors called for an end to the emergency law and requested a multiparty system and competitive elections (Hinnebusch 2012: 103). The critics gained the attention of the old regime members (the conservatives) who had still been in possession of power in the military and security apparatus. They approached Bashar and warned him of the diminishing impacts of the regime's power to open up the society. Thereafter, the regime's leadership went on a counteroffensive. This social and political reform period came to a rapid end after the first seven or eight months of Assad's presidency. With the re-imprisonment of a number of prominent pro-democracy activists, it was apparent that the regime reverted to administrative and repressive methods (Lesch 2012: 9; Perthes 2004: 17). The appearance of truly independent political associations and leadership figures was suppressed by the government's use of long-standing emergency powers (Leverett 2006: 37).

On the side of media, experts on Syria, such as Volker Perthes or Alan George, argued that the reforms on the Syrian media could not have created press freedom, as any periodical and even non-political advertising magazines required a licence from the prime minister. The editors were forced to distribute their magazines through the official public sector distributing agency, while journalists were punished for spreading false news, ranging from fines to three years' imprisonment (Perthes 2004: 21; George 2003: 121). The Press Law of 2001 banned many subjects relating to "national security, national unity, details of the security and safety of the army its movements, weapons, supplies, equipments and camps" (George 2003: 121). The press has remained under the tight control of the regime, and therefore, the modernised authoritarianism of Assad can be seen in the media field as well.

Despite these attempts at suppression, the regime could not totally suppress the nascent civil society. Civil society "committees" continued to meet and voice their opinions through the Lebanese press, international media and online (Perthes 2004: 19). For instance, the Civil Society Movement was one of these civil society organisations, which appeared following the death of Hafez Al-Assad and worked for the democratisation of Syria. It was subsequently suppressed by the regime in 2001, although the leaders of this movement claimed, "the civil society movement had not failed. On the contrary, it had encouraged the creation of a broad alliance of opposition tendencies" (George 2003: 63). Moreover, it was apparent that the reforms fortified the techno-savvy younger generation. Although the very strict Press Law of 2001 imposed threatening limitations on media, Syrians pushed past the "red lines" on the Internet. Like other activists around the world, they learned to use proxies which enabled them to access blocked websites through other servers or through the use of service providers in Lebanon or Jordan (Shaery-Eisenlohr 2011: 4). Increased activism was apparent in the offline space as well. After the fall of Baghdad in 2003, disillusioned youth groups from the activist milieu of anti-war demonstrations emerged. These groups intended to rebuild society from below by fighting corruption, raising awareness and encouraging people to take the initiative. Yet, none of these groups was in direct opposition to the regime and did not explicitly propagate any anti-regime ideas; they merely sought to work separately from the government (Magout 2012). Thus, although the modernisation attempts of Bashar al-Assad activated certain segments of Syrian society, there was no evidence of a direct challenge by these groups to the regime until the Syrian uprising in March 2011. According to Michel Kilo, a prominent Syrian writer, human rights activist and one of the central figures of the "Damascus Spring", civil society had yet to crystallise in Syria and the local society was busy with its own concerns, afraid of politics and distanced itself from the political sphere (Kilo 2011: 161). This will become crucial to the argument advanced in this book.

Egyptian Islamists and Secular Intellectuals

In Egypt, civil society was constituted by various actors, but competition for political power was mostly polarised between two main camps: the Islamists and secular intellectuals (Abdelrahman 2007: 25). This polarisation formed the major characteristics of the civil society in Egypt. During the Mubarak period, along with these two main groups, NGOs and think tanks also emerged to influence the debate about political and social power (McGann 2008: 31). Like Bashar al-Assad, Mubarak was tolerant towards these groups early in his presidential tenure; he enjoyed a honeymoon period of goodwill and a broad range of political support. As soon as he arrived, he made an emotional impact on the formal political scene and released 31 of the main political detainees imprisoned during the Sadat period (McDermott 1988: 74). He also brought opposition parties into discussions and changed the electoral laws in 1987. With the February 1987 electoral law, independents obtained the chance to win seats in the parliament. The proliferation of newspapers also fortified the sense of freedom on a scale which had not been felt for some years (McDermott 1988: 74).

This reform period, however, did not last long. During the 1990s, the regime's crackdown on the Muslim Brotherhood signalled a change in the state's approach to civil society. The crackdown was justified by the violent activities of the Islamist group, Al Islamiyya (Cook 2012: 165; Hazmawy and Grebowski 2010). Although the Muslim Brotherhood had not engaged in any violent activities for decades, Mubarak and the people around him used this violence to demonise the Muslim Brotherhood, claiming the Brotherhood to be the intellectual source behind the violence that the country was confronting (Cook 2012: 165).

The continuing presence of the emergency laws also proved the reluctance of the Mubarak regime to promote freedom in civil society. The laws allowed the regime to suppress any political dissent and break up any public gatherings or demonstrations against it (McGann 2008: 31). The state of emergency thus further consolidated the president's absolute authority. It empowered the prime minister and minister of the interior to constrain the movement of individuals, search persons or places without warrants, tap telephones, monitor and ban publications, forbid meetings and detain suspects without trial. Even gatherings of more than five people were illegal under these laws (Lesch 2011).

If the press structure of Egypt was analysed, it would be apparent that in parallel with the liberalisation overtures of the Mubarak regime, the press enjoyed a period of significant freedom in the first half of the 2000s (Black 2008: 2). Cook (2012: 196) describes this period as a renaissance of the Egyptian press, with privately owned journals appearing and journalists beginning to ask questions, editorialise and break stories that were previously off-limits, such as questioning the very nature of the Egyptian state and society. Privately owned journals *Al-Fajr* and *Al-Dostour* were edited by these independent journalists who reflected the common disaffection with government and authority by using vibrant and entertaining journalistic techniques (Black 2008: 3). One of the most important newspapers, *Al Masry al Youm*, which played an important role in paving the way to the Egyptian revolution, also emerged during this period. Founded in 2004 by human rights activist and media entrepreneur, *Al Masry al Youm*, it became known for its serious and investigative journalism and its opposition-friendly editorial line, informing the public about official corruption, political malfeasance and foreign policy in a dispassionate manner (Cook 2012: 196). The leftist newspaper *Al Badeel* was another important journal in this liberation period. This publication was launched in 2007 and edited by the prominent journalist and intellectual, Mohamed Al-Sayed Saeed (Saleh 2009).

This liberalisation of the press did not last long. By 2007, relations began between the government and press deteriorated. According to Black (2008: 5), this was the result of general unease in the society, with a succession of debates on executive power (Black 2008: 5). For Mubarak, who was 79 years old in 2007 and had no apparent successor other than his second son, Gamal, it was not the best time to be subjected to severe criticism ("Mubarak's mission" 2009). The deterioration proceeded with the arrests of journalists, such as Ibrahim Eissa, editor of the *Al-Dustour* newspaper. Three days before Eisaa's trial, three staff at the opposition daily, *Al-Wafd*, were also sentenced to prison for writing articles about members of the judiciary (Black 2008: 5). These custodial sentences for journalists were not a new procedure. The modern structure of state control over news media in Egypt was shaped during the state's centralised, socialist phase (Black 2008: 5). Despite the high point of press freedom at the beginning of the 2000s, Mubarak did not change the structure and legal framework of the Egyptian press. The primary instrument of state control was the Supreme Press Council which, led by the Al Ahram establishment, owns the national press and holds the right to issue licences to the rest. Along with the national press, whose editors were largely self-censoring, mindful of their payments, there was also an imported media, known as the "Cyprus Press". The issues of the Cyprus Press were passed through the Ministry of Information, which notified the state if the media violated certain "red lines" and were able to issue threats to stop certain stories going to press (Cooper 2008: 4). The Press Law, on the other hand, paved the way for the imprisonment or fining of journalists for insulting the president or top government officials, negative portrayals of the armed forces, or for criticism of the leadership of a foreign country (Amin 2014: 129). Although the Press Law of 2006 removed some of the articles of the Penal Code, it still contained provisions for custodial sentences for vague and easily abused offences (Black 2008: 7).

Overall, with the appearance of the privately owned newspapers, the media became relatively free in Egypt in comparison to Syria (El-Ghobashy 2011: 22). Nevertheless, as with all aspects of public life, it was still under the strict scrutiny of the regime, which acted decisively whenever an editor was perceived as a threat. Control was not restricted to the censorship of cultural and media productions. It extended to the operation of labour unions, which was closely checked. Lesch (2011) cites as an example public sector workers who were required to join a union under the authority of the Egyptian Trade Union Federation. Ironically, the Trade Union Federation was directed by NDP officials who were often businessmen themselves (Lesch 2011: 36–37). Hence, even professional syndicates were brought under government control. The strict control by the regime over different segments of civil society showed similarly to that of Bashar al-Assad; the liberalisation narrative of Mubarak could easily take a patriarchal and authoritarian turn when its authority was threatened.

With this degree of state control exerted in the political sphere, the Internet arose as the sole prominent medium that widened the bottom-up struggle against authoritarianism and prevented full control of ideas, values and information (Sadiki 2009). Digital media enabled people to meet and discuss beyond the ears of the secret police, but also helped the public to experience things directly (Cole 2014: 66). Recent research of Valeriani and Vaccari (2015) also reveals that accidental exposure to online information increases as interest in politics decreases. Egypt's active and diverse blogosphere in particular played a significant role in maintaining the political engagement of middle-class Egyptians with less interest in politics.

Egypt had a large array of blogs, some written by a group of core bloggers, such as Wael Abbas, Hossam El-Hamalawy, Nora Younis, Alaa and Mona Abdel Fattah and Mahmoud Salem (known as Sand Monkey), who started using blogs during the experimentation phase (Radsch 2008). In the year leading up to the revolution, these blogs played different roles. First of all, they were a great information source for a public who only heard about torture through rumours and had no personal experience of it (Cole 2014: 66). It was also an important source for journalists, who reported on the torture being routinely carried out by the police. Al Jazeera journalist Howayda Taha, for example, produced a documentary in 2007 following up on stories uncovered by bloggers relating to torture in Egyptian police stations (Nanabhay and Farmanfarmaian 2011: 76). One of the interviewees for this book, blogger Wael Abbas,[2] was one of those who offered important sources to the media by covering the repression and torture of the Egyptian police in his blog, Misr Digital (Egyptian Awareness).

Coverage of police corruption and torture had a directly damaging effect on the regime's reputation (Cole 2014: 67). After all, while the Syrian regime was known as a security state, its Egyptian counterpart had long been known as a police state. Coverage reinforced reputations and expectations. The term *police state* not only appeared because of the coercive apparatus routinely used to detain and torture people, but also given that the police were the chief administrative arm of the state, taking on the roles of several agencies (El-Ghobashy 2011). Yet, the police were also particularly associated with bribery and torture. In 2006, for instance, 21-year-old microbus driver Emad El-Kabir was arrested while trying to stop a police assault on his cousin. During his detention, the police officers who interrogated El-Kabir tortured, beat and raped him (Faris 2013: 6). A video of this ferocious attack on El-Kabir was made by one of the officers using a cell phone and sent to others. Wael Abbas posted the video of El-Kabir being tortured on his blog and accelerated the transition of information (Faris 2013: 6). Until 2007, this widespread torture by the police department received little press attention in Egypt. However, with the arrival of the mobile web, videos of torture went viral thanks to the efforts of bloggers like Abbas. The courageous work of such bloggers filled an important hole in Egyptian civil society and gave voice to the oppressed. Nevertheless, as Faris (2013) argues, the elite blogs soon became hubs. Hubs were crucial to provision, routing, co-ordination and dissemination of information that increased the ease and efficiency of navigating the network, yet their popularity made it complicated for new voices to gain any kind of traction in the Egyptian blogosphere (Faris 2013: 6). This in turn prevented other Egyptians from achieving similar results and becoming heard.

Although the blogs could not give the same portion of opportunities to each individual, they were crucial in building networks and virtual communities on the Internet (Herrera 2014: 14). These online communities enabled interaction between members of the opposite sex, people of different religions and different ideologies. Scholars who analysed the Egyptian blogs such as Herrera (2014) and Cole (2014) suggest that exchanging views and acknowledging other identities helped transform the youth culture in the years leading up to 2011 (Cole 2014). The young would gain familiarity with diverse opinions. Discovering their similarities, some of the youth bloggers who supported Muslim Brotherhood, for instance, deserted to liberal and leftist groups or joined the Muslim Wasat (Centre) party (Cole 2014: 71). Bloggers would also play an important role in improving the repertoires of contention of Egyptians; their significant role will be explained in detail in the next section, on the repertoire of contention of Egypt.

REPERTOIRES OF CONTENTION

Syria—The Kingdom of Silence

In the absence of a direct political challenge, Bashar al-Assad succeeded in maintaining a stable regime over the 10 years of his presidency and believed in himself and his regime with growing self-confidence. David W. Lesch who conducted many interviews with Bashar al-Assad over his presidency claimed that Assad felt safe, secure and popular in the country and was beyond condemnation when the Arab Spring began (Lesch 2012: 48). Thus, he was shocked when the uprising in the Arab world started to penetrate his own country in March 2011. What particularly led Assad and his supporters to believe that the regime could weather the storm of the Arab Spring was primarily the lack of opposition to the regime.

Because of Syria's turbulent political development following independence, Syrians generally preferred to distance themselves from engaging in political activities that might risk instability. Looking across their borders towards Lebanon and Iraq, two countries ethnically and religiously similar to Syria, Syrians could see how political disorder could violently destroy the fabric of society (Lesch 2012: 50). Moreover, the West was considered a threat and a colonising aggressor by modern Arabs. The state's discourse always drew attention to the critical nature of the current moment. Enemies always stood at the gates and threatened the safety of the homeland. Describing the current situation as critical and alarming helped the state justify the continuation of the emergency law (Saleh 2003: 64). Hence, due to the country's vulnerability to instability, according to Syrian blogger Otrakji, the vast majority of society was tolerant of the regime, even if they impatiently expected reforms (Nabki 2011).

However, the desire for stability was only one of the factors that fostered disdain of political life amongst Syrians. All ideologies in Syria enforced by the regime have played an active role in creating public distance from political life and transforming the state into the active centre of initiative (Saleh 2003: 62). This reinforced the understanding that the centralisation of power was crucial to the speeding up of the modernisation process and justified civil society's exclusion from engagement in the country's political and intellectual life.

Hence the total presence of the state could be felt in every aspect of life and social interactions. It could be claimed that this divergence of civil society from political life created "a kingdom of silence" (Al Hendi 2012). The mechanisms of the reproduction of these ideologies span the institutions, including the school system, NGOs, popular organisations and the media (most notably, TV). The strategic narratives, which were used to create a shared meaning of the past, present and future of international politics were effective in shaping the behaviours of domestic and international actors (Miskimmon et al. 2013: 2). The narratives of Syrian leaders showed great sensibility with regard to all that was external and the obvious presence of an inner/outer dichotomy. This dichotomy was used to implement values quite antagonistic to change, paternalistic in nature, and linking everything, from the people to the state (Saleh 2003: 62).

According to Lisa Weeden (1999), who explored state domination in the country during the Hafez Al-Assad period, another important factor that distanced the society from politics was the regime's attempt to create a "cult of personality" around the personality of the president. In her influential book, *Ambiguities of Domination*, Weeden argues that the official rhetoric in Syria often included absurd statements regarding the personality of Assad, for instance representing him as a knight or the father of the country. She then questions whether this rhetoric could produce belief or emotional commitment amongst citizens (Weeden 1999: 506). As Weeden finds that people did not believe these "mystifications" that the regime put, she further asks: "Why would a regime spend scarce resources on a cult whose rituals of obeisance are transparently phony?" (Weeden 2002: 723). Weeden's ethnographic work and interviews in Syria help us to understand the significance of creating a cult of personality: Assad's personality cult was not derived from the legitimacy, charisma or hegemony that would have brought support for Assad and his policies. Yet, the symbolic production was important not only to exemplify but also produce power for the regime. This empty official rhetoric and slogans cluttered the public space and drained citizens' political energy. Insinuating these symbols into people's daily life via credible threats of punishment habituated people to perform the gestures and pronounce slogans reflecting their obedience.

Again, according to Weeden (1999), the semiotic discourses also produced the principle of social auto-totality, which refers to the people as encouraging and enforcing obedience in each other. As examples, Weeden shows shopkeepers who posted official iconography in their windows, even though they were not required to (Weeden 1999: 10–12). These mechanisms of control established during the Hafez Al-Assad period could also be seen during the term of Bashar al-Assad.

Personal initiatives by regime fanatics were also helpful in spreading this propaganda. For instance, in one of his speeches, a senior Ba'ath Party official referred to Bashar as the father of the country, which recalled the Hafiz Al-Assad period. Regime supporters also displayed pictures of the Trinity of Hafez, Bashar and his brother, Hafez, suggesting that the hereditary rule should continue over the coming years (Crisis Group 2011: 7). The only difference between Bashar al-Assad and Hafez Al-Assad periods was how they described the cult of personality around the personality of Bashar. Unlike his father, the Syrian media portrayed Bashar as a celebrity, a loving family man, and pictured him with a map or flag of Syria, which emphasised his attachment to his country (Magout 2012: 6). The regime could thus continue to apply its cultural mechanisms of control during the Bashar al-Assad period as well.

However, the country's rapidly changed media ecology made control of this rhetoric difficult. In the 1990s and 2000s, the Internet and satellite TV channels took up an informative role in Syria and created a much more interactive and engaging space. While the satellite TV channels helped in the emergence of what Lynch (2007) calls a "new Arab public", civil society activists used the Internet as part of the political project to resist the Syrian regime (Lynch 2007; Shaery-Eisenlohr 2011). Shaery-Eisenlohr illustrates this with reference to the online magazine of Tahara, which reviews scholarship, culture and literature on women's issues and informs the society about women's problems (Shaery-Eisenlohr 2011). In the same way, Noueihed and Warren (2012: 48) cite TV programmes such as "Al-Itijah Al-Muakis" or "the Opposite Direction", which played a significant role in eroding the "cult of personality" that Arab dictators had carefully tried to construct. The programme was famous for asking provocative questions about the Arab dictators and for comparing the Arab World with the West. The effects of this change in the media landscape were also felt in Syria. Anderson (2013), who performed ethnographic fieldwork in Aleppo between 2008 and 2009, found that the cultural mechanism of the state's control was no longer very effective in Syria. People who had had enough of the corruption, patriarchal and family-based system adopted a narrative of scorn and lament. They not only mocked or castigated the other but also drew attention to his or her failings. Thus, the society was well aware of the problems that were ongoing in the country (Anderson 2013: 454).

And yet despite the rising political awareness in the country, there was not any direct opposition attempt against the regime until the start of the Syrian uprising in 2011. As the editors of the online magazine Tahara expressed, the society could publish information about the problems they experienced but could not act on these due to the heavy hand of the state (Shaery-Eisenlohr 2011). The Syrian interviewees of this book indicate that the Syrian people obeyed the norms and did not at first approve of the protests, due to a lack of protest experience and, secondly, due to enforcement and coercion structures that have operated in the Syrian security state. When they compared themselves to the Egyptian or Tunisian protesters, the interviewees claimed that what actually differentiated them from other protesters was the lack of political experience of the Syrian activists, who were afraid to be punished. Caroline,[3] for instance, a Christian Syrian living in Damascus and detained by the Assad regime in 2012 for her civic engagement, mentioned that:

"Before the revolution in Egypt, people were allowed to gather, had political parties; people were exposed to political life. In Syria, we were away from politics. We were raised in Syria and our parents used to tell us that we shouldn't talk with anyone about our religion or about politics" (Aslan 2015). Thus, as a reason for youth political inactivity, Caroline cites the limited political space in Syria and the lack of repertoires that would draw on social ties and past protest experience.

Syrian interviewees also claimed that they were afraid to be arrested by the security services that were present in every district of Syria. "The Mukhabarat", meaning the security services in Syria, were well known for their repressive activities. They often referred to soft use of power, such as pre-emptive fear and intimidation to deter potential unrest and disruptive activities by real or perceived opposition elements (Lesch 2012: 65). Apart from the few Syrians who had been politically active over the past years, such as the members of the "Civil Society Movement", there was a lack of experienced community in Syria that was capable of organising a solidly structured movement.

Egypt's Wave of Protests

While Syria was described as the Kingdom of Silence, not a day passed in Egypt without several incidents of collective action (Abdelrahman 2014: 29). The waves of protests and uprisings date back to 1952. From Nasser to Mubarak, groups such as university students, state security officers, industrial workers, judges and small farmers have challenged the state and its repressive and exclusionary policies (Abdelrahman 2014: 29). Nevertheless, the years of the 2000s had particular importance in Egyptian protest history. Scholar El-Ghobashy (2011) describes Egypt's streets in the 2000s as a space full of parliaments, negotiating tables and battlegrounds rolled into one (El-Ghobashy 2011).

What differentiated the 2000s from previous periods was the start of the second Palestinian uprising or Intifada (Gunning and Baron 2013: 38; Abdelrahman 2014). This was not the first time that Egyptians protested in support of the Palestinians. During the 1980s and 1990s, groups such as the Muslim Brotherhood, leftists and Arab nationalists had joined in the protests in support of the Palestinians. However, none of these protests could create a sustained wave (Gunning and Baron 2013: 37). Mubarak's rule weakened the capacity of street activism and the appearance of the dirty war in the 1990s between the regime and

Islamist militants resulted in the death of street activism. Consequently, during the 1990s, Egyptians were as afraid to engage in collective action as Syrians. Egyptian journalist and blogger, El-Hamalawy (2011), for instance, mentions that one could only whisper the Mubarak's name in the 1990s.

Nevertheless, the second Palestinian Intifada gave birth to street activism once again. As in Syria, it intensified the demands of the protesters and drove the growth of groups and networks behind them. The Egyptian Popular Committee for Solidarity was the main organising body behind the movement and consisted of Arab nationalists, Marxists, Islamists, civil society and professional organisations, intellectuals, artists and public figures (Abdelrahman 2014). The emergence of the Popular Committee showed that the solidarity of various groups during the 2011 Egyptian protests was not formed over 18 days, but rather dates back to the beginning of the 2000s. The Committee later organised many more public demonstrations in mosques, Egyptian universities and schools in support of the Palestinians. It later became an exemplar for future groups, such as the Popular Committee Opposing US Aggression against Iraq, which brought together the initiative of secular civil society organisations and leftist activists (Abdelrahman 2014: 31).

These protests also served to reconnect the long-term activists of the 1970s, who had become fragmented and gone in different directions due to different ideological trends (Gunning and Baron 2013: 42). Along with establishing national networks, the Palestinian Intifada was crucial to creating links amongst Egyptian and global activists. Egyptian groups such as the Popular Committee would work in co-ordination with global groups under the umbrella of an anti-war movement and the Global Justice Movement (Abdelrahman 2014: 34).

Another important attribute of the second Palestinian Intifada in Egypt was its rising anti-regime dimension during 2002. When the USA invaded Iraq, the protests would grow dramatically and the anger exploded on an even larger scale (El-Hamalawy 2011). Inspired by the Seattle activists, on 20 March 2003, approximately 2000–3000 people occupied Tahrir Square for a day. Unlike the Syrian state, which supported the protests against the foreign powers, the response of the Egyptian regime to the protesters was harsh. During the March demonstrations, hundreds were arrested and beaten. This amplified social discontent and protesters started to link the events in Iraq with the behaviour of the regime itself. Slogans such as "Leave Mubarak, leave" filled the streets (Gunning and Baron 2013: 39). The movement not only criticised Mubarak's authoritarian regime, but also questioned the foundation of the post-1952 state and its destruction at the hands of Mubarak and his allies (Abdelrahman 2014: 30). Hence, unlike the Syrians protesters, Egyptians articulated out loud their discontent with the regime from the early 2000s. They were also subjected to harsh repression, which changed the risk perception of the protesters. While the protesters were prepared to risk more to protect what they had achieved, the regime, in turn, had to increase repression to achieve its goals (Gunning and Baron 2013: 45).

The news also sparked the motivations for political protest and shifted the political culture in the country. As El-Hamalawy states (2011), scenes aired by Al Jazeera and other satellite television stations of the Palestinian revolt or the US war in Iraq deepened the tensions and gradually smashed the wall of fear amongst the Egyptian protesters. The Palestinian Intifada also marked the beginning of Internet activism in Egypt, which took the form of email lists and yahoo chat groups. Boycott lists, pictures of dead and atrocities circulated via email and Yahoo groups (Faris 2013: 90). Hence, when the 2011 Egyptian protests erupted, Egyptians activists already had 10 years of experience on the Internet.

Yet, what particularly shifted the repertoire of contention in Egypt was the Kefaya (which means "enough" in Arabic) movement. The Kefaya movement is considered a landmark of the Egyptian protest wave for having been the first movement in Egypt that directly targeted the president and his family. It began as a sustained campaign for democratic change in 2004 and, along with the Mubarak family, it targeted the parliamentary elections in 2005. The two main slogans of the protesters, which broke the taboo against criticising Mubarak and his family, were: "no to a new mandate, no to inheritance of the presidency" (El-Hamalawy 2011; Abdelrahman 2014: 37). Co-ordinated by a steering committee, Kefaya was formed of different political groups, including Al Karama, Al Wasat, the Revolutionary Socialists, Al-Ghad and many independent figures. The movement also benefitted from the experience of the protest groups, who had already been present in the Palestinian Intifada and anti-Iraq War protests (Gunning and Baron 2013: 42). These experienced activists introduced new forms of opposition politics and different tactics to a significant number of young people who had not previously taken part in politics (Abdelrahman 2014: 37).

For instance, they launched new initiatives in Egypt, such as the organising of petitions or creation of stay at home protests. They also used sweeping tactics, which involved masses of people sweeping brooms to demonstrate the need to clean up the country's politics (El-Hamalawy 2011).

The innovative tactics of Kefaya activists and the violent response of the police towards the protesters would motivate many and gave way to the appearance of new protest groups, such as Shayfeen (We can see you) and El-Shari Lena (the Street Is Ours). The Street Is Ours campaign, for instance, was launched by women and activists who had been abused by the police and sexually assaulted on the "Black Wednesday" of 2005, constitutional referendum day. The repression by regime loyalists on that day would trigger such sentiments as anger and anxiety amongst the protesters and gave way to the emergence of the Street Is Ours campaign on 9 June 2005. This campaign was intended to collect testimonies from women and men on the violence and harassment they faced in their daily lives (Abdelrahman 2014: 38; "This day in history" 2005).

Kefaya also led to the emergence of numerous factional movements, such as the March 9 Movement, Students for Change, Doctors for Change, Youth for Change, Workers for Change, Journalists for Change and the Judges Club Protesters (Gunning and Baron 2013: 38). Amongst these, Youth for Change (Al Shabab min Ahl Al Tagheer) played a crucial role in the nascent repertoire of contention of the Egyptians. By applying a novel tactic, called a guerrilla strategy by Azimi (2005), members of the group targeted the working-class Cairo community. Groups of between two and four activists visited public squares or parks and engaged in discussion with people on the street. To get the attention of the public and link their daily concerns with the failure of the political process, they created street shows, such as the erection of small exhibitions with flyers or the performance of street theatre (Azimi 2005).

These social and mainstream media activities by the Kefaya activists were effective in altering the political and the nascent blog culture in the country (El-Hamalawy 2011). Some of the activists, such as Manal and Alaa Abd El Fattah, created an Internet hub for Egyptian bloggers, motivating others to engage in "citizen journalism". The bloggers sent pictures of demonstrations, but also those of police abuses, and posted them on the web as a contemporary archive (Azimi 2005). As a prominent blogger, Wael Abbas[4] started to cover what was happening in the streets with the Kefaya Movement as well. During my interview with him, he states:

> I was active online before 2004, but active in the streets from 2004. I took my camera and got to the streets. I was joining the demonstrations and posting the videos online.

Hence, unlike the 1990s, the rallies in 2004 were not only covered by journalists from major government dailies, such as Al-Akhbar, Al-Gumhuriya and Al Ahram (Faris 2013: 89), but were also framed by prominent bloggers like Wael Abbas, who helped in the dissemination of news coming from protesters to outsiders.

These bloggers would play an important role during the clash over judicial rights in 2005–2007. The tension rose in 2005 with a public struggle between the regime, which wanted to rig the presidential elections for the NDP, and the judiciary, which was constitutionally tasked with overseeing the fairness and legality of the elections themselves (Faris 2013: 92). As a result of the clashes, the Judge Club divided into two parts. One was formed by judges who believed they should not take part in elections that would likely be fraudulent, while the other thought they should exercise their power and took into consideration the limits placed on judicial freedom (Faris 2013: 92). During the elections, two judges who claimed the elections had been rigged; hundreds of activists and leaders from the Kefaya movement and Muslim Brotherhood were arrested. Leading Arabic-language bloggers such as Wael Abbas informed the world about the protests that broke out after the arrests (Faris 2013: 92). In our interview, Wael[5] describes these protests as the most active years of the Egyptian bloggers prior to the 2011 Revolution:

> The protests lasted for two weeks, lots of activists who took part in the Judge protests were bloggers and most of us were arrested. I was wanted back then but I succeeded in escaping. They sent approximately 50 people to jail, mostly blogger activists. Bloggers spent two months in jail. I know two of them who were sexually assaulted. So things became quiet afterwards.

The words of Wael demonstrate the active presence of the Egyptian bloggers in the online and offline protest spheres before 2011. The bloggers posted the dates and times of the protests, joining the demonstrations and experiencing police brutality themselves along with other protesters in attendance. Consequently, their participation on the ground enabled them to witness the events and spread the word swiftly as they occurred. They also developed their repertoire of contention and got used to risking more in order to achieve their aims.

In 2006, the silent segment of the society, the labourers, also joined the protest wave that was rising in the country. They created groups of opposition that would constitute the overwhelming majority of strikes between 2006 and 2009 in Egypt (Gunning and Baron 2013: 59; Schimidinger 2014: 134). The first big labourers' protest in Egypt took place in Mahalla El-Kubra, whose name would crop up again and again in the following years. Mahalla sits to the north of Cairo and is known for its public sector textile and clothing manufacturing (Agbarieh-Zahalka 2008). The protests started when the workers of Mahalla El-Kubra were not given two months of incentive pay promised by Prime Minister Ahmed Nazif (El Sharnoubi 2013). After a week of complaining and organising small protests, 24,000 labourers began to protest on 7 December 2006. Encouraged by the female workers who first went to strike, more than 10,000 others occupied the factory over the following three days and formed a strike committee to organise the logistics of the strike, speak to the media and negotiate with the authorities. On the fourth day, government officials, who did not want the protests to spread to other sectors, offered a 45-day bonus to workers (El-Mahdi 2011: 387). This strike generated a wave of workers' protests across Egypt, "crossing different sectors of the economy and industries—from Mahalla, to Kafr al-Dawwar, to Shibin al-Kum; from spinning and weaving to cement, to the railways, the metro and public transport workers" (Bassiouny and Said 2008). The labourers used different repertoires of contention during these protests, ranging from sit-ins to factory occupations and street protests (Bassiouny and Said 2008).

When textile workers called for a new strike in 2008 against rising food prices, activists from outside Mahalla el-Kubra also offered support. One of these was an El Ghad Party functionary, Esraa Abdel Fattah,[6] who would later become very well known as "Facebook girl". Esraa started to use Facebook groups for connecting with people after watching a film called Heya Fawda ('it is chaos'). Co-directed by Egyptian directors Youssef Shahin and Khaled Youssef, the movie was released in cinemas in 2007 and was about the long-standing problem of police corruption and brutality (Amin 2014). In my interview with Esraa,[7] she mentions that the movie deeply influenced her and she thought everyone from each town and city of Egypt should see it. In pursuit of this aim, she created a Facebook group with the name of the movie and asked everyone to join. She describes this stage:

> At that time there were no Facebook pages and there was a very small number of Facebook users. So only 500 people joined the group that I founded. Then, we went to the cinema to watch the film. There were approximately 100 people in the salon. Everyone knew each other. When the movie finished, we started to shout "down with Mubarak". We went out into the streets and encouraged Khaled Youssef with our slogans to produce many more films that give the same message. It was a great event for that time.

Hence, locally produced films such as Heya Fawda played a crucial role in spreading anger and frustration amongst Egyptians before the 2011 protests. Like Heya Fawda, the big budget film, "Yacoubian Building", by Alaa Aswany, was another wake-up call to many Egyptians. The film was released in Egypt during 2006 and its shocking scenes demonstrated the reality of societal ills, including the corruption, sexual repression and religious fundamentalism plaguing contemporary Egyptian society (Amin 2014). Along with many Egyptians, the key figures of the 25 January 2011 protests such as Esraa were influenced by these films and started to become politically active for change in Egypt.

What brought particular fame to Esraa was her second Facebook group, launched in 2008 and dedicated to the April 6 labour protests. Esraa[8] describes the appearance of this Facebook group:

> Ahmed Maher (one of the co-founders of the April 6 youth movement) asked me to create a new Facebook group to understand the reactions of people, whether they wanted to join in the labour protests or not. I told Ahmed this is not about a movie – we need to brainstorm and decide on the photo of the group, the name of the group and the description to make the strike attractive. So we had a lot of discussions with Ahmed Maher and finally called the Facebook group, 'April 6'. We said that it was a general strike in Egypt. We even had a competition amongst people and asked them to make a profile picture for the group. Lots of people participated in this competition. We then voted and chose one of the photos.

The words of Esraa Abdel Fattah shows that she and Ahmed Maher automatically adopted a horizontal infrastructure for the organisation of protests and asked members to indicate their views about the symbols of the movements, such as the picture for the page, which would later be a raised fist. The members also shared their views about how to spread the idea of strikes to Egyptians. Esraa mentions that one of the members proposed that the group to write "April 6 is a general strike" on currency notes after they got their salaries. To spend this money would spread the word. Thus, this communication by the April 6 Facebook group members helped to improve Egyptian protesters' repertoire of contention. The April 6 Facebook group was also the first time that Facebook was used as a co-ordination tool in Egyptian protests.

The group turned the localised labour movement into an international event and triggered an international cascade (Faris 2013: 97). Within two weeks, 7000 people joined the page. Considering the small number of Facebook users in Egypt (790,140 Facebook users) in 2008, this was a big success (Faris 2008: 2). Yet, despite all these preparations and support from people, the protesters could not initiate large protests in Cairo streets. There could be two reasons behind the lack of protests in the capital. First of all, the organisers offered different protest options to the participants. Esraa[9] reports these options as:

> We first of all told them: 'do not go to your work and instead go to demonstrations in the street. If you cannot go to the demonstration, just stay at home and do not go to work. If you do go to work, just raise your Egyptian flag or wear black clothes. If you cannot do any of these, do not buy anything from the shops'. I also called people who have shops and asked them to close their shops at that time. So we offered people different ways to participate in the protests.

Thousands of Egyptians heeded the call and stayed home on that day (El Bahr 2011). Secondly, the heavy presence of the police all across Egypt on 6 April might have been the factor that prevented the spread of the protests in Cairo's streets. Esraa, for instance, says that she was the co-ordinator of the Tahrir Square protest and was charged with collecting people around the streets of Tahrir and bringing them to the square. In Tahrir, KFC fast food restaurant was the protesters' meeting point. Yet, when Esraa came to Tahrir with her group, she realised that there were no staff in KFC, and instead there were policemen. She suggests that when they were talking on the phone, their phones were certainly monitored. Thus, although the Internet was significant in getting the attention of people and internationalising a local event, it also alerted the regime itself. When the protesters came to the protest site on 6 April, the police were ready for the protests and quickly arrested those who came to the square. Esraa indicates that they could only walk along two streets and she was then arrested, along with all the people with her.

On the other hand, the scene in the north of the country, in the city of Mahalla El-Kubra, was quite different by comparison to what was going on in Cairo (Faris 2013: 98). Mamdouh Salah,[10] a protester from Mahalla and participant in labour movements from 2006, describes the protests in Mahalla as a "warzone". The police were also well-prepared for the strike in Mahalla El-Kubra. To prevent the mobilisation of workers, the textile factory of Mahalla was surrounded by police. The heavy presence of the police prevented workers of the factory from joining the protests. Nevertheless, despite the absence of workers, the citizens of Mahalla El-Kubra succeeded in turning the main square of Mahalla, Al-Shown Square, into a protest site. In response, the police used live ammunition and rubber bullets that resulted in the death of a 15-year-old bystander and wounded more than 100 people ("Egypt: Protestors killed marking revolution" 2011). Thus, all these protest methods flourished on the ground or on Facebook were key to improving the repertoire of contention of the Egyptians. The protests also taught them how to confront police during protests.

The protests in Mahalla El-Kubra, which erupted, again from time to time over the next two years, were also key to improving the capacities of grass-roots organisations in the city. These grass-roots organisations created dense social networks that generated shared emotions and a collective identity within the labour community. The significance of grass-roots organisations and collective identities would become more apparent a year later, when the 6 April organisers tried to follow up their success with an anniversary strike. In the absence of the grass-roots organisation, protests on the anniversary were mostly a resounding failure (Faris 2013: 102). The shared grievances which brought workers around a collective action a year previously were also absent during the 2009 protests. Esraa Abdel Fattah[11] mentions that what brought success to the 2008 protests was the approach of the organisers to labourers. She claims:

> In 2008, we first asked people to join us for political reasons but we realised they did not listen to us so we decided to go to them. We needed to choose issues which made it easy for us to approach people. This was the reason for the success of this group at this time. People found that we did not talk about our issues, but rather talked about their issues.

In line with Esraa's words, Mamdouh, a protester from Mahalla El-Kubra, mentioned that in order to mobilise the labourers, one should first understand their ideology and then approach them with these messages. Mamdouh notes that:

> If you do not know the problems of labourers, you cannot reach them, because labourers want to do their work and go home afterwards to rest. They do not have time to do lot of meetings. So from 2006 to 2011, labourers, taxi drivers, people in the villages, farmers and students came together and talked about the problems of labourers.

Mamdouh's words prove that the protests which emerged in Egypt before 2011 were significant in teaching the Egyptian activists how to mobilise people. Citizens acquired organisational experience while they tried to force unresponsive officials to enact or revoke specific policies (El-Ghobashy 2011: 24). The following chapters of this book will show that the online and offline experiences of the activists were key for the success of the 2011 Egyptian protests.

The next section briefly summarises the similar and differing variables that had an impact on the appearance of protests in both countries, based on the historical contexts analysed up until 2011.

Factors That Triggered the Collective Action in Egypt and Syria

Similarities

1. One of the major factors that gave way to the uprising in Egypt and Syria was a similar problem that was observed in other Arab Spring countries as well: **A crumbling economy**. Conditions in Egypt reached rock bottom in the last year of the Mubarak period: Poverty, disease, unemployment, lack of health and deteriorating education could all be listed as daily problems for most Egyptians

(Al Aswany 2011: 5). In the new neoliberal system in Egypt, areas such as labour rights, rural livelihoods, progressive taxation, social security and pension schemes were all targeted and all operated against the interests of regular citizens (Abdelrahman 2014: 12). Consequently, the public started to realise that this new order was restructured to their disadvantage. The economic outlook was no better in Syria. When Bashar al-Assad came to power, he realised the necessity of improving the economic structure and initiated economic reforms during the first decade of his rule. To liberalise the economy, the regime established private banks in 2004 as part of a series of monetary reforms. Moreover, the long-awaited stock exchange was finally introduced in 2009 and steps were taken to make Syria investor-friendly in order to attract foreign investment (Lesch 2012: 60). However, according to Lesch (2012: 60) and many other experts, such as Perthes (2004), making changes at the margins would not improve the economy overall and Assad could or would not undertake the systematic changes that had been long awaited in the society. The economic problems and deficiencies indigenous to Syria, as well as the economic downturn of late 2008, led the Syrian economy to collapse in early 2011 (Lesch 2012: 62). Thus, similarly to Egypt, the market economy of Syria had been in free-fall for a long time and the Syrian public had gradually lost patience.

2. Another problem was **economic inequality** in both countries. Although the new neoliberal order of Egypt brought an increase in consumer expenditure per capita between 2000 and 2008, only a privileged few had enjoyed this prosperity. While the *nouveau rich* were enjoying high-end shopping centres, gated communities and private jets, the population below the poverty line reached 22 per cent in 2006[12] (Cook 2012: 175). The high level of corruption in Egypt served to entrench this gap. In 2008, Transparency International gave Egypt a score of 2.8 out of 10 on its corruption index; a state is considered corrupt when it receives a score lower than 5 (Soliman 2011: 30). The pervasive corruption was also major problem of the Syrian political system. Instead of resolving this corruption, the Assad regime helped to enrich the privileged few who possessed political or family connections with the regime. As in the reign of Hafez Al-Assad, the regime has also continued to maintain an alliance with big businesses at the expense of smaller

businesses and without regard to Syrian society, which suffered the most from diminishing government support (Haddad 2012). Thanks to pervasive corruption and privileged access to wealth, a *nouveau rich* emerged under the presidency of Assad. Lesch (2012: 64) claims that due to the monopolisation of important sectors in a few hands, the anger of the protesters was directed towards "the economic oligarchs" such as Rami Makhlouf (the cousin of Bashar al-Assad and the principal owner of Syriatel), as much as towards the president. In 2011, these economic inequalities played a major part in triggering people's anger.

3. Another similarity between Egypt and Syria was the **large youth population** in both countries: more than 50 per cent of the population in Syria are indicated as under the age of 24[13] by CIA World Factbook. This has led to pressures on schools and universities, as well as on the labour market (Perthes 2004: 29). According to 2012 data,[14] unemployment in Syria was estimated to be 18 per cent and this was higher amongst the younger generation. The rising number of young people in the Egyptian society was also a key problem. There were 45 million Egyptians under 35 years old. The official unemployment level for the under-30 age group was 21 per cent and this number almost doubled the overall total (Osman and Samei 2012: 216). During the 2011 protests, this pool of young people would play an important role in mobilising the public (Gunning and Baron 2013: 43).

4. A fourth similarity is that both Syrian and Egyptian **protesters were inspired by protests in other countries**. The political, social and economic factors cited above have been ongoing problems of Syrian and Egyptian regime. As a major factor that triggered the Syrian uprising, the protests that broke out in Tunisia and Egypt could be cited, where lifetime-mandate rulers also reign and where the protesters were acting in response to similar problems to those the Syrians had long been faced with. As the popular revolts in Tunisia and Egypt threw off the yoke of repression, Syrians became hopeful for their future. The interviewees for this study specifically pointed out that the Egyptian protesters inspired them. For instance, Amer[15] expresses the significance of Egypt for Syria in saying that: "Ben Ali left Tunisia, and we heard about this. But with the news from Egypt, most Arabs were following the situation day-by-day, because Egypt is different from other countries in the

Arab region". Amer's response proves how important and inspiring the situation in Egypt was for the Syrian people. When I asked him whether he felt these feelings when Mubarak was brought down, he said "yes, for sure, there was a similar experience before and you believe that yours will be similarly successful". By taking Amer's answer into consideration, one can easily argue that the success stories from Egypt that spread via TV or over the Internet sparked hope for change in Syria. On the other hand in Egypt, the activists have been already planning to protest on the police day, January 25, against the police brutality. Yet when I asked my Egyptian interviewees what brought millions to the streets, they all mention the Tunisian effect. They indicated that they thought Tunisians were not better than Egyptians so they could do it as well.[16]

5. Social movement scholars such as Castells (2012), Jasper (2011) and Melucci (1996) have long emphasised the importance of **emotional investment** on the part of participants in collective action. The factors above show that emotional investment had been present in Syria and Egypt for a long time. The interaction of the Internet and mainstream media helped to increase and reveal this accumulated emotional investment. For instance, as the images of Egyptian or Tunisian protesters appeared on their televisions or Internet pages, Syrian interviewees for this study claimed that they became hopeful about their future and felt excited about possible changes in the country. In today's movements, the horizontal communication enabled by information and communication technologies is the fastest, most autonomous, interactive, reprogrammable and self-expanding means of communication. As the communication process is faster and more interactive today, it will more easily propel enthusiasm and motivate people towards collective action (Castell 2012: 15). However, the empirical chapters, which follow this chapter, will reveal that the transformation of emotion into action is a complicated process, and does not only depend on the communication process.

Differences

1. One major difference between Egypt and Syria was the **sectarian tendencies of the Syrian government**. As was already explained at the beginning of this section, sectarian tendencies

by the Syrian minority government have always been problematic for the regime's stability. Although Assad tried to accommodate the interests and sensibilities of urban Sunnis to prevent their opposition to his regime and co-opt the key actors in the Sunni community, sectarian divisions were reinforced by ongoing trends in Syria. These divisions intensified the Islamisation of the Sunni majority. Moreover, as urbanisation has advanced since 1982, the urban class has had a favoured position over rural Sunnis (Leverett 2006: 36). According to Leverett (2006: 36), this gave way to a social climate even more encouraging to Islamist resurgence than that of the early 1980s. In line with Leverett's (2006) arguments, an interviewee for this research, Hediye Levent[17] notes that when the uprising started, it was easy for journalists to observe these Islamised groups, which gathered in mosques and took to the streets together, chanting against the regime and the Alawis. Although these Islamist groups would not affect the ignition phase of the peaceful protests and the unity of Christians and Muslims at the start of the protests, it would have an impact on the further stages of the uprising.

2. As is claimed above, **Egypt had a more embedded and sophisticated repertoire of contention than Syria**. The short-term factors that appeared in 2010 further improved the repertoire of protesters in the country and developed the political consciousness of the Egyptian citizens. The return of ElBaradei, the former director of the International Atomic Energy and Nobel Prize winner, to Egypt, initiated this enlightenment period. Once he returned to his home country, ElBaradei galvanised support for change by unifying the forces of past movements. The online and offline campaigns of ElBaradei and its impacts on the society will be covered in more detail in Chapter 3.

3. Another significant event that increased the political awareness of the Egyptian public was the **2010 National Assembly elections**. The parliamentary elections were significant for both the regime and opposition, as this was the first step towards the fateful presidential elections of 2011. It was crucial for the NDP to win an overwhelming majority in parliament, since this would make it easier for the party to choose a new political option to succeed Mubarak, which was expected to be Gamal Mubarak (Ghanem and Mustafa 2011: 400). To achieve victory, the regime exerted

pressure on opposition candidates and imposed restrictions on the press and other mass media. Hence, it limited the public visibility of the opposition, and especially the Muslim Brotherhood, during the 2010 elections. The succession campaign for Gamal Mubarak inspired reactions from many segments, including the army, elites and even regular Egyptians inside Egypt. After all, in contrast to the Syrians, Egyptians had never accepted the idea of a monarchical republic in which the son inherits the father's throne (Al Aswany 2011: 5). Thus, the 2010 parliamentary elections were held in an atmosphere of suspicion. The victory of the NDP brought with it reactions from different segments in the society.

4. Unlike Syria, **public opinion in Egypt was also inflamed by two major domestic events**. The first was the beating and death of Egyptian citizen Khaled Mohamed Saeed at the hands of police in Alexandria in June 2010. The horrific photos of Khaled were widely distributed to the public via the Facebook page of Wael Ghonim under the moniker, "We are all Khaled Said". The widespread use of torture and corruption by the security services had been captured and screened by bloggers such as Wael Abbas for a long time. Yet, unlike blogs, the Facebook page also offered the chance to create an interactive space where Ghonim galvanised the public around shared grievances against police brutality. The page soon became a focal point for the distribution of information about the protests and for activating the public (The Facebook page and its impact will be analysed in more detail in the next chapter). The second factor that triggered the public's anger was the Alexandria church bombing during a New Year's Eve service at the Al-Qiddissin Church, when a bomb went off in the street outside, killing 21 people and injuring 70. The suspected suicide attack drew attention to the lack of security in Egypt's streets. Many Christians also become cynical towards the regime and considered the notion that the attack might in fact have been planned by the regime to trigger sectarian tension in the society. All of these incidents triggered anger towards the police who, despite having draconian powers, repeatedly proved incapable of anticipating and thwarting such violence ("Egypt bomb kills 21" 2011). For instance M.S,[18] an Egyptian activist from Cairo, states that:

The Egyptian population is conservative. I am not speaking in terms of religion but with regard to risk-taking. They prefer security but because of what happened in the church bombing, they realised that even minimum security and sustenance is not there. So why would we not resist this regime? The only thing that the regime promised to provide was not there anymore.

Hence, prior to 2011, Egyptians began to realise that the state was not providing security to a wide range of social groups to which it had once committed itself. This furthered the anger against the regime and generated frustration within the society. (Abdelrahman 2014: 12)

Conclusion

Comparing the structural factors that are unique to Egypt and Syria shows that the Syrian president may not have enjoyed the same level of centralised executive authority as Egyptian President Mubarak did, but he had gradually changed the opposing views around him and created more centralised power for himself. Economically, both Egypt and Syria moved towards a market economy, but the free market philosophy would not be the solution to the economic problems and corrupt systems of either. On the media side, although civil society and the Egyptian media enjoyed far more freedom than Syria, liberty in Egypt also had its limits. When Mubarak's authority was challenged, his liberalisation narrative quickly took a patriarchal and authoritarian turn. The most striking difference amongst the factors analysed in this chapter was between the repertoire of contention in each country. While Egyptians became more and more used to engage in protest activities, most Syrians were still afraid to engage in any activity against the regime. Hence, the Syrians did not have the motivation or ability to participate in politics.

Nevertheless, despite the reluctance of Syrians to participate in politics, in 2011 both countries witnessed the emergence of the biggest anti-regime protests in their history. When analysing the factors that triggered the Syrian and Egyptian protests, this chapter showed that in both countries new political opportunities appeared with reforms that gave way to an opening up in the society and laid the groundwork for an uprising. The collapsing economy and high level of repression in the countries played an important role in the accumulation and explosion of emotions. Thus, the factor, accumulation of emotion, that have long been emphasised by social movement scholars as vital to the formation of collective action have been present in Syria and in Egypt, and neither the Arab Spring nor the Internet was the only reasons behind the appearance of the protests. However, both cases prove that external factors, such as the broader Arab Spring protests, and the presence of information and communication technologies, can be cited as new political opportunities which helped to further motivate public action.

The differences between the Egyptian and Syrian regimes did not affect the appearance of collective action in both countries. Nevertheless, the different political contexts had cumulative impacts on the progression and outcomes of both protest movements. For instance, there was no significant division in the rank of the big businesses of Syria. As with the relatives of Assad, other major players coming from different families also remained loyal to the regime (Haddad 2012). In Egypt, however, there was growing discontent amongst the military and elites towards the rigged elections in favour of the NDP. Although the Egyptian military did not play a big role in the protests, it also did little to quash the protests and hence changed the destiny of Egypt. Political opportunities, such as the arrival of ElBaradei and the failure of the police to protect citizens, further differentiate the political scenes in Syria and Egypt. The existence of political figures such as Baradei and the performances of Egyptian protesters had a positive impact on the online and offline activism.

Comparing Egypt and Syria provides us with the opportunity to test the role of new media technologies in the protests in these two countries, one with experienced activists, a developed civil society and a multiparty system and the other without any of these structural features. I deal with the impact of these structural factors on the use of new media technologies in far greater detail in the next chapters.

Notes

1. http://www.internetworldstats.com/me/sy.htm.
2. Interview with Wael Abbas, Egyptian blogger, Cairo, 2014.
3. Skype Interview with Caroline, Syrian activist, 2013.
4. Interview with Wael Abbas, Egyptian blogger, Cairo, 2014.
5. Interview with Wael Abbas, Egyptian blogger, Cairo, 2014.
6. Interview with Esraa Abdel Fattah, Egyptian activist and blogger, Cairo, 2014.
7. Interview with Esraa Abdel Fattah, Egyptian activist and blogger, Cairo, 2014.
8. Interview with Esraa Abdel Fattah, Egyptian activist and blogger, Cairo, 2014.

9. Interview with Esraa Abdel Fattah, Egyptian activist and blogger, Cairo, 2014.
10. Skype Interview with Mamdouh Salah Eldeen, Egyptian activist, 2014.
11. Interview with Esraa Abdel Fattah, Egyptian activist and blogger, Cairo, 2014.
12. https://www.cia.gov/library/publications/the-world-factbook/geos/eg.html.
13. https://www.cia.gov/library/publications/the-world-factbook/geos/sy.html.
14. https://www.cia.gov/library/publications/the-world-factbook/geos/sy.html.
15. Interview with Amer Assayed Omar, Syrian journalist and activist, Istanbul, 2013.
16. Interview with Sheriff Alaa, Egyptian activist, London, 2013.
17. Skype interview with Hediye Levent, journalist, 2013.
18. Interview with M.S (he preferred to be mentioned with his initials), Egyptian activist, Cairo, 2011 and 2014.

References

Abboud, S. (2015). Locating the "social" in the social market economy. In R. Hinnebusch & T. Zintl (Eds.), *Syria from reform to revolt: Volume 1: Political economy and international relations.* Syracuse, NY: Syracuse University Press.

Abdelrahman, M. (2007). The politics of 'uncivil' society in Egypt. *Review of African Political Economy, 29*(91), 21–35. https://doi.org/10.1080/03056240208704582.

Abdelrahman, M. (2014). *Egypt's long revolution: Protest movements and uprisings.* New York, NY: Routledge.

Agbarieh-Zahalka, A. (2008, May–June). Egyptian workers impose a new agenda. *Challenge Magazine.* Retrieved from http://www.challenge-mag.com/en/article__212.

Al Aswany, A. (2011). *On the state of Egypt: What made the revolution inevitable.* New York, NY: Vintage Books.

Al Hendi, A. (2012, October). The kingdom of silence and humiliation. *Foreign Policy.* Retrieved from http://foreignpolicy.com/2012/10/16/the-kingdom-of-silence-and-humiliation/.

Amin, S. (2014, July 8). Did culture and arts prepare Egypt for revolution? *Index on Censorship.* Retrieved from https://www.indexoncensorship.org/2014/07/culture-arts-egypt-revolution/.

Anderson, L. (2013, August 7). The politics of scorn in Syria and the agency of narrated involvement. *Journal of the Royal Anthropological Institute, 19*(3): 463–481. https://doi.org/10.1111/1467-9655.12045.

Aslan, B. (2015). The mobilisation process of Syria's activists: The symbiotic relationship between the use of information and communication technologies and the political culture. *International Journal of Communication, 9,* 2507–2525.

Azimi, N. (2005, September 1). Egypt's youth have had enough. *OpenDemocracy.* Retrieved from https://www.opendemocracy.net/democracy-protest/enough_2794.jsp.

Bassiouny, M., & Said, O. (2008). A new workers' movement: The strike wave of 2007. *International Socialism, 118.* Retrieved from http://isj.org.uk/a-new-workers-movement-the-strike-wave-of-2007/.

Black, J. (2008, January). Egypt's press: More free, still fettered. *Arab, Media & Society, 4,* 1–13. Retrieved from http://www.arabmediasociety.com/articles/downloads/20080114230358_AMS4_Jeff_Black.pdf.

Castells, M. (2012). *Networks of outrage and hope: Social movements in the Internet age.* Cambridge, UK: Polity Press.

Cole, J. (2014). *The new Arabs: How the millennial generation is changing the Middle East.* New York, NY: Simon & Schuster.

Cook, S. A. (2012). *The struggle for Egypt: From Nasser to Tahrir Square.* New York, NY: Oxford University Press.

Cooper, K. J. (2008, September). Politics and priorities: Inside the Egyptian press. *Arab, Media & Society, 6,* 1–9. Retrieved from http://www.arabmediasociety.com/articles/downloads/20080929131720_AMS6_Ken_Cooper.pdf.

Crisis Group. (2011). *Popular protest in North Africa and the Middle East (VI): The Syrian people's slow motion revolution* (Middle East & North Africa Report No. 108). Retrieved from http://www.crisisgroup.org/~/media/Files/Middle%20East%20North%20Africa/Iraq%20Syria%20Lebanon/Syria/108-%20Popular%20Protest%20in%20North%20Africa%20and%20the%20Middle%20East%20VI%20-%20The%20Syrian%20Peoples%20Slow-motion%20Revolution.pdf.

Dam, N. V. (1996). *The struggle for power in Syria.* New York, NY: Palgrave Macmillan.

Egypt bomb kills 21 at Alexandria Coptic Church. (2011, January 1). BBC. Retrieved from http://www.bbc.co.uk/news/world-middle-east-12101748.

Egypt: Protestors killed marking revolution. (2015, January 26). Human Rights Watch. Retrieved from http://www.hrw.org/news/2015/01/26/egypt-protesters-killed-marking-revolution.

El Bahr, S. (2011, November 3–9). Ahmed Maher: Action for a new Egypt. *Al-Ahram Weekly.* Retrieved from http://weekly.ahram.org.eg/2011/1071/profile.htm.

El-Ghobashy, M. (2011). *The Praxis of the Egyptian Revolution.* Middle East Report 258, pp. 1–7. Retrieved from http://www.merip.org/mer/mer258/praxis-egyptian-revolution.

El-Hamalawy, H. (2011, March 2). Egypt's revolution has been 10 years in the making. *The Guardian.* Retrieved from http://www.theguardian.com/commentisfree/2011/mar/02/egypt-revolution-mubarak-wall-of-fear.

El-Mahdi, R. (2011). Labour protests in Egypt: Causes and meanings. *Review of African Political Economy, 38*(129), 387–402. https://doi.org/10.1080/03056244.2011.598342.

El Sharnoubi, O. (2013, April 6). Revolutionary history relived: The Mahalla strike of 6 April 2008. *Ahram Online.* Retrieved from http://english.ahram.org.eg/NewsContent/1/0/68543/Egypt/0/Revolutionary-history-relived-The-Mahalla-strike-o.aspx.

Faris, D. (2008, September). Revolutions without revolutionaries? Network theory, Facebook, and the Egyptian blogosphere. *Arab Media and Society.* Retrieved from http://www.arifyildirim.com/ilt508/david.faris.pdf.

Faris, D. (2013). *Dissent and revolution in the digital age.* New York, NY: Palgrave Macmillan.

George, A. (2003). *Syria: Neither bread nor freedom.* New York, NY: Zed Books.

Ghanem, A., & Mustafa, M. (2011). Strategies of electoral participation by Islamic movements: The Muslim Brotherhood and parliamentary elections in Egypt and Jordan, November 2010. *Contemporary Politics, 17*(4), 393–409. https://doi.org/10.1080/13569775.2011.619763.

Gunning, J., & Baron, I. Z. (2013). *Why occupy a square: People, protests and movements in Egyptian revolution.* New York, NY: Oxford University Press.

Haddad, B. (2012). Syria, the Arab uprisings, and the political economy of authoritarian resilience. *Interface, 4*(1), 113–130.

Hazmawy, A., & Grebowski, S. (2010). From violence to moderation. *Carnegie Papers.* Retrieved from http://carnegieendowment.org/files/Hamzawy-Grebowski-EN.pdf.

Herrera, L. (2014). *Revolution in the age of social media.* London, UK: Verso.

Hinnebusch, R. (2001). *Syria: Revolution from above.* New York, NY: Routledge.

Hinnebusch, R. (2009). Teaching & learning guide for: Modern Syrian politics. *History Compass, 7*(6), 1616–1622.

Hinnebusch, R. (2012). Syria: From 'authoritarian upgrading' to revolution? *International Affairs, 88*(1), 95–113. https://doi.org/10.1111/j.1468-2346.2012.01059.x.

Hopwood, D. (1992). *Egypt 1945–1990: Politics and society.* London, UK: Routledge.

Jasper, J. M. (2011). Emotions and social movements: Twenty years of theory and research. *Annual Review of Sociology, 37,* 285–303.

Kilo, M. (2011). Syria... The road to where? *Contemporary Arab Affairs, 4*(4), 431–444. https://doi.org/10.1080/17550912.2011.630884.

Kramer, G. (1992). *Liberalization and democracy in the Arab world*. Middle East Report 174: Democracy in the Arab world, 22. Retrieved from http://www.merip.org/mer/mer174/liberalization-democratization-arab-world.

Lesch, A. M. (2011). Egypt's spring: Causes of the revolution. *Middle East Policy, 18*(3), 35–48. https://doi.org/10.1111/j.1475-4967.2011.00496.x.

Lesch, D. W. (2012). *Syria: The fall of the house of Assad*. Cornwall, UK: TJ International.

Leverett, F. (2006). *Inheriting Syria: Bashar's trial by fire*. Washington, DC: Brookings Institution Press.

Lynch, M. (2007, Spring). Blogging the new Arab public. *Arab Media & Society, 1*. Retrieved from http://www.arabmediasociety.com/?article=10.

Magout, M. (2012, June). *Cultural dynamics in the Syrian uprising*. Graduate Section of the British Society for Middle Eastern Studies Annual Conference. London: The London School of Economics and Political Science.

Marsot, A. L. A. (2007). *A history of Egypt: From the Arab conquest to the present*. New York, NY: Cambridge University Press.

McDermott, A. (1988). *Egypt from Nasser to Mubarak: A flawed revolution*. New York, NY: Croom Helm.

McGann, J. G. (2008, August). Pushback against NGOs in Egypt. *The International Journal of Not-for-Profit Law, 10*(4). Retrieved from http://www.icnl.org/research/journal/vol10iss4/special_3.htm.

Melucci, A. (1996). *Challenging code: Collective action in the information age*. Cambridge, UK: Cambridge University Press.

Miskimmon, A., O'Loughlin, B., & Roselle, L. (2013). *Strategic narratives, communication power and the new world order*. New York, NY: Routledge.

Mubarak's mission. (2009, August 19). *The Economist*. Retrieved from http://www.economist.com/node/14252582?zid=309&ah=80dcf288b8561b012f-603b9fd9577f0e.

Nabki, Q. (2011, May 2). *Talking about a revolution: An interview with Camille Otrakji* [Web log post]. Retrieved from http://qifanabki.com/2011/05/02/camille-otrakji-syria-protests/.

Nanabhay, M., & Farmanfarmaian, R. (2011). From spectacle to spectacular: How physical space, social media and mainstream broadcast amplified the public sphere in Egypt's 'revolution'. *The Journal of North African Studies, 16*(4), 573–603. https://doi.org/10.1080/13629387.2011.639562.

Nassif, H. B. (2013). Wedded to Mubarak: The second careers and financial rewards of Egypt's military elite, 1981–2011. *The Middle East Journal, 67*(4), 509–530.

Noueihed, L., & Warren, A. (2012). *The battle for the Arab Spring: Revolution, counter-revolution, and the making of a new era*. New Haven, CT: Yale University Press.

Osman, A., & Samei, M. A. (2012). The media and the making of the 2011 Egyptian revolution. *Global Media Journal, 2*(1), 1–19. Retrieved from http://www.db-thueringen.de/servlets/DerivateServlet/Derivate-25453/GMJ3_Samei_final.pdf.

Perthes, V. (1992). *Syria's parliamentary elections: Remodelling Asad's political base*. Middle East Report 174: Democracy in the Arab World, 22. Retrieved from http://www.merip.org/mer/mer174.

Perthes, V. (2004). *Syria under Bashar al Assad: Modernisation and the limits of change*. London, UK: Oxford University Press.

Pfeifer, K. (2012). Economic reform and privatization in Egypt. In J. Sowers & C. Toensing (Eds.), *The journey to Tahrir: Revolution, protest and social change in Egypt*. London, UK: Verso.

Radsch, C. (2008, September). Core to commonplace: The evolution of Egypt's blogosphere. *Arab Media and Society*, 6. Retrieved from https://www.researchgate.net/publication/237494130_Core_to_Commonplace_The_evolution_of_Egypt%27s_blogosphere.

Sadiki, L. (2009). *Rethinking Arab democratization: Elections without democracy*. Oxford, UK: Oxford University Press.

Saleh, Y. (2003, March). The political culture of modern Syria: Its formation, structure & interactions. *Conflict Studies Research Centre*, 57–68. Retrieved from http://www.mafhoum.com/press7/225P9.pdf.

Saleh, Y. (2009, April 9). Al-Badil newspaper turns weekly. *Daily News Egypt*. Retrieved from http://www.dailynewsegypt.com/2009/04/09/al-badil-newspaper-turns-weekly/.

Schimidinger, T. (2014). The role of labour unions in the transition and political struggles after Mubarak. In A. Hamed (Ed.), *Revolution as a process: The case of the Egyptian uprising* (pp. 127–155). Bremen, Germany: Academic Press.

Seale, P. (1998). *Asad: The struggle for the Middle East*. London, UK: California Press.

Shaery-Eisenlohr, R. (2011). From subjects to citizens? Civil society and the Internet in Syria. *Middle East Critique, 20*(2), 127–138. https://doi.org/10.1080/19436149.2011.572410.

Sika, N. (2012). Youth political engagement in Egypt: From abstention to uprising. *British Journal of Middle Eastern Studies, 39*(2), 181–199. https://doi.org/10.1080/13530194.2012.709700.

Soliman, S. (2011). *The autumn of dictatorship: Fiscal crisis and political change in Egypt under Mubarak*. Stanford, CA: Stanford University Press.

Stacher, J. (2012). *Adaptable autocrats: Regime power in Egypt and Syria*. Stanford, CA: Stanford University Press.

This day in history. 25 May 2005: Mubarak thugs sexually assault journalist Nawal Ali. (2013, May 25). *Egypt Independent*. Retrieved from http://www.egyptindependent.com/news/day-history-25-may-2005-mubarak-thugs-sexually-assault-journalist-nawal-ali.

Valeriani, A., & Vaccari, C. (2015). Accidental exposure to politics on social media as online participation equalizer in Germany, Italy, and the United Kingdom. *New Media and Society, 18*(9), 1857–1874.

Weeden, L. (1999). *Ambiguities of domination: Politics, rhetoric, and symbols in contemporary Syria.* Chicago, IL: University of Chicago Press.

Weeden, L. (2002). Conceptualizing culture: Possibilities for political science. *The American Political Science Review, 96*(4), 713–728.

Zisser, E. (2001). Does Bashar al-Assad rule Syria? *Middle East Quarterly, 10*(1), 15–23. Retrieved from http://www.meforum.org/517/does-bashar-al-assad-rule-syria.

Three Styles of Leadership in the Egyptian Protests

"Even from England, it was obvious that change was coming to Egypt" said Mona (pseudonym),[1] sitting in a cafe in Zamalek, an upscale district in Cairo. Mona was living abroad until she decided to join the ElBaradei campaign, the National Association for Change (NAC), in 2010. After joining, she first organised the NAC's London conference and then returned to her home country to contribute to the change. She was in the Revolutionary Youth Coalition (RYC) that organised the 2011 Egyptian protests on the ground. Like Mona, many of my Egyptian interviewees started to be politically active long before the 2011 protests and took key roles in the ElBaradei campaign.

As Tarrow claims, movement organisations in a country progress by interacting with cultural artefacts, power holders and other movements (2011: 127). Like organisation, leadership skills have to be learned. The education and trial and error experiences of activists are crucial to teaching leadership skills as a movement unfolds (Oberschall 1973: 58). As explained in Chapter 2, several incidents of collective action had been taking place daily in Egypt since 1952 (Abdelrahman 2014). Egypt has also enjoyed multi-party presidential elections since 2005. Young Egyptians acquired leadership skills during these past protests and election campaigns. Although the elections in Egypt were not as democratic as they may have seemed, they paved the way for the rise of different political figures, such as ElBaradei, who would bring hope for change

© The Author(s) 2020
B. Aslan Ozgul, *Leading Protests in the Digital Age*,
Palgrave Studies in Young People and Politics,
https://doi.org/10.1007/978-3-030-25450-6_3

to Egyptians in 2010. The activists who worked for the ElBaradei campaign acquired organisational capabilities in both the online and offline spheres. Thanks to these past collective actions, the Egyptian activists had already created loose ties amongst each other. In this chapter, I will explain the impact of the ElBaradei campaign on the activities of the Egyptian activists in 2011 and discuss how the political structure in Egypt and the past protest experiences of the Egyptian activists were crucial to enabling them to co-ordinate during the preparation phase.

Moreover, this chapter analyses the organisational structure of the 2011 Egyptian protests in both the online and offline spheres by giving voice to the key activists who took part in the organisation of the 2011 Egyptian protests. The uprisings in Egypt were first described as mostly enacted by spontaneous, largely leaderless, multimodal networks (Castells 2012: 56; Bennett et al. 2014). Without the intense effort, time and money of the organisers, a movement's messages travel from person to person in cyberspace via personalised networks and spread the emotions of hope and excitement amongst the public (Castells 2012). Further analysis by Paolo Gerbaudo (2012: 145) showed that protests are not merely propelled by the self-motivated sharing of ideas, plans and images on social media channels but that leaders still have a presence in digitally networked protests. Contrary to conventional leaders, they apply a soft leadership through their use of the interactive and participatory medium of the Internet (Gerbaudo 2012: 145). More recently, Poell et al. (2015) further developed this theory of soft leadership. Analysing the interaction between the administrators and users of the "Kullena Khaled Said" ("We are all Khaled Said") Facebook page, the most popular online platform during the Egyptian revolution of early 2011, they argue that the page administrators should be understood as connective leaders. They fostered user participation and connected users in online communication streams and networks by employing sophisticated marketing strategies.

In this chapter, I develop these theories of communication scholars and demonstrate the specific contributions of different types of leaders to the organisation of the Egyptian protests. The interviews, conducted with the administrators of the Kullena Khaled Said Facebook page (KKSFP) and the experienced activists on the ground will show that there were three sorts of communicators in the Egyptian protests, namely: *hybrid* leaders, who are both communicators on the ground and in the digital sphere; *soft* leaders, who communicate online and spread the message about the offline organisation; and *experienced* leaders on the ground. The analysis also reveals that the political structure of Egypt and the repertoire of leaders determined their capacity to organise the protests, both online and offline.

The Rise of Hope and the Beginning of a Change

Egyptians' excitement for change started long before the Tunisian revolution and even before the launch of the KKSFP. The protests that led to the Egyptian revolution are mostly associated with this Facebook page, which soon became the biggest dissident Facebook page in Egypt, with 473,000 users (Preston 2011). It facilitated the assembly of the cosmopolitan youth around a collective identity and played an important role by being one of the most important organisational tools of the 2011 Egyptian protests (Gerbaudo 2012: 48). Nevertheless, what primarily brought hope for change and motivated different groups to unify around a unique aim was neither the launch of the KKSFP nor the Tunisian protests. It was, rather, the vow of Mohamed ElBaradei to fight for change. As the fourth Egyptian ever to win the Nobel Prize, in 2005, ElBaradei was widely admired as an influential figure in Egypt (Kirkpatrick 2012) and held the top position at the International Atomic Energy Agency (IAEA), as its Director General. When his third term came to an end in 2009, he decided to return to his country. In a statement to the Egyptian newspapers, he indicated his discontent with the way Egypt was being governed (Ghonim 2012: 40). The speech encouraged many young Egyptians inside and outside the country that had previously not had much interest in politics. For instance, John Albert,[2] from the organising team of the 2011 Egyptian protests in Alexandria, started to become politically active after listening ElBaradei's speech on TV. He described this period:

> I was at my uncle's house. I saw ElBaradei on TV and everyone was talking about him. I asked people who ElBaradei was. Every TV channel that I turned on was saying that he was a spy. I started to read about him and watched his speeches on TV. That's how he caught me. He was talking about very logical things. His ideas were making sense and I decided to join his campaign.

John's words reflect how ElBaradei attracted the attention of the youth. Despite being the subject of heavy media criticism, ElBaradei motivated young Egyptians like John to be politically active. Wael Ghonim, the founder and admin of the KKSFP, defined the significance of this return:

> We all craved for an alternative. We needed a saviour and we were ready to pour our hopes onto any reasonable candidates. Finally, two years after the April 6 Movement began, Egyptian activists believed they had found one. (Ghonim 2012: 40)

Here, Wael points to the difference between ElBaradei's campaign and the past movements in Egypt's history. Before ElBaradei's return, Egypt had already witnessed significant movements such as the Palestinian intifada and the April 6 movement in the 2000s. As explained in Chapter 2, these protests were crucial in reconnecting the long-term activists of the 1970s with each other, and in the formation of a new stratum of young activists (for full discussion please see Chapter 2). However, ElBaradei's arrival was the first time Egyptians had felt there was an alternative to Mubarak for the presidency.

Although ElBaradei did not mention his willingness to run in the presidential elections, Egyptians started to consider him as an alternative to Mubarak (Haulohner 2010). Even before his arrival, a campaign was launched for his return. Hundreds of Egyptians rallied via opposition newspapers and a Facebook page called "ElBaradei, President of Egypt 2011". The aim of the campaign was to welcome ElBaradei on his arrival at the airport and show the support of Egyptians for him (Haulohner 2010).

Esraa Abdel Fattah,[3] who took part in the organisation of the campaign on the ground, explained that millions gave support to the Facebook page. Nevertheless, they only targeted 5000 people, as it was difficult to control more than 5000 people on the ground. Esraa[4] said:

> With other groups, we were more than 10 people who co-ordinated the venue because it was a reception at the airport. People came from different governments and we co-ordinated whether they came with buses or via other channels.

For young Egyptians like Esraa, joining the organisation of the ElBaradei campaign was a prodigious experience that taught them how to co-ordinate different groups from different provinces in both the online and offline spheres. Esraa's experience, gained from taking part in the April 6 movement, meant that she had realistic expectations about the number of Facebook followers who would attend the protests. She expected only 5000 people from the 1 million page followers to arrive at the venue. Esraa[5] states that the activists learned to assemble people around a unique aim during two stages. The first stage was the April 6 movement. The second started with the emergence of ElBaradei. I believe Esraa referred to these two movements in particular, and not others in the past, as she took part in both the April 6 movement and the ElBaradei campaign as an organiser. She believes organising these two movements helped her and other organisers to acquire relevant and effective communication skills. As Tarrow (2011: 29) explains, "societal movements are repositories of knowledge of particular routines in a society's knowledge which helps them to overcome the deficits in resources and communications typically found among disorganised people". My Egyptian interviewees indicated that the collective actions organised through the Internet prior to the 2011 Egyptian protests helped them to understand how they could overcome the deficits in resources and communication during the digitally supported collective action.

The venue organised for ElBaradei's arrival in 2010 was the first mobilisation experience that the Egyptian activists acquired during the ElBaradei campaign. The size of the venue convinced ElBaradei that the youth was eager to bring change into the country. He responded to the demands for change by launching the "National Association for Change" (NAC), a coalition including important media and political figures, such as former presidential candidate Ayman Nour and the media veteran Hamdy Kandil. Some leaders of the Muslim Brotherhood, political parties such as the liberal Democratic Front, the left-wing Nasserist Al Karama and moderate Islamist Al Wasat party and the organisations, the Revolutionary Socialists, Egyptian Women for Change and the April 6 Youth Movement, were also amongst this group (Ghonim 2012: 44). Unlike traditional party programmes, the NAC developed a seven-point pamphlet that requested legal and constitutional changes including free and fair elections, an independent judiciary and a free press in Egypt (Ghonim 2012: 45).

Although the NAC did not articulate a long-term objective, it was crucial to create political awareness amongst Egyptians and show the public the problems caused by the electoral system. John,[6] who actively worked on the ground level and online part of the campaign, said:

The National Association for Change started at the end of 2009. At that time, people were so scared to talk about politics. We had papers with us explaining seven demands for change. We also explained to people the election process such as "your ID should be with you", "the elections should be pure". We tried to make people care about politics. We hoped that this would be the right way to choose the president. So, we went to the streets, the supermarkets and everywhere and tried to convince people to sign these seven demands.

John's words demonstrate that the organisers of the 2011 protests had started to get in touch with the public and pushed Egyptians to talk about politics on the street before 2011. The ElBaradei campaign helped Egyptians to overcome the spiral of silence in the country before the 2011 protests, and it became an important stage in developing the public's own consciousness.

The campaign of ElBaradei was also effective in creating a new group of admins, including the administrators of the KKSFP, Abdelrahman Mansour, Wael Ghonim and Sami (pseudonym). Wael Ghonim, for instance, stated that although he first hesitated in signing the petition, two days later he entered all his personal information. His fear turned into excitement and he realised that he had started a new phase in his life by being publicly opposed to the state (Ghonim 2012: 46). The names cited on the petition reinforced the network of trust and solidarity amongst people. The campaign thus helped to break the culture of fear and activated important figures that would play a crucial role in the 2011 Egyptian protests.

One month after ElBaradei's arrival, in March 2010, the Facebook page had 82,069 supporters, compared to Gamal Mubarak's Facebook account with 6583. ElBaradei's Twitter account also became the most followed account in Egypt over a short period of time (Ghonim 2012: 48). Like the Syrian Revolution 2011 Facebook page explained in this Chapter, the activities of the admins, together with traditional media, contributed to the page's popularity.

However, despite its online success, the NAC could not make a political breakthrough. ElBaradei declared that he would not run for presidential elections unless the constitution was changed to permit fair competition from independent candidates (Siperco 2010). In addition, in spite of the high number of "likes" of the ElBaradei page, the campaign resulted in a limited number of signatures. In his book Ghonim

(2012: 50) relates this failure to the media, public opinion changed in accordance with the media campaign launched by the state. The state authorities had banned ElBaradei from appearing in the Egyptian media prior to the 2010 parliamentary elections ("Elections in Egypt" 2010). Private media channels that had previously covered news pertaining to him also began to keep their distance following his return to Egypt in 2010. Thus, the media that had once brought fame to the social media accounts of ElBaradei ended up taking attention away from him.

THE EXPERIENCED ACTIVISTS

Despite its failure to bring about political change, the ElBaradei campaign had an important impact on the activists who operated in the online or offline spheres during the 2011 Egyptian protests. Many members of the Revolutionary Youth Coalition (RYC), the organising body of the protests in Tahrir Square, had previously been member of the ElBaradei campaign. Along with ElBaradei campaign members, the RYC was composed of representatives of the April 6 movement such as Ahmed Maher, Asmaa Mahfouz, the Muslim Brotherhood youth, and Kullena Khaled Said page administrators like Wael Ghonim (interview with Mona). The RYC was also supported by Kefaya, the Tomorrow Party, the youth wing of the liberal Democratic Front party, the Freedom and Justice Movement of the Muslim Brotherhood, the leftist popular Democratic movement for Change (Hasd), the Free People's Front, and the Nasserist Hamdeen Sabahi's Karam party youth wing (Cole 2014: 146). Hence, the Revolutionary Youth Coalition's members obtained leadership experience as the movements unfolded in Egypt from the 2000s. The group was composed of the representatives of political movements, organisations and party members that were once working separately (Shehata 2011). Although they have never claimed to be the leaders of the movement, they acted as the leaders and planned the times and places of the marches that took place during the 2011 Egyptian protests. Ahmed Maher, for instance, co-founder of the April 6 movement, met with the leaders of the soccer fun Ultras from Ahly and Zamalek, which would galvanise crowds during the protests (Ghonim 2012: 145). Hence, this was an organisation with extensive national networks, which helped the group to exchange information and pool resources for the protests (Della Porta and Diani 2006: 125).

The RYC decided on the starting points and methods of co-ordination for the protests. In his memoir, Ghonim, the administrator of KKSFP, writes that "I was not a field expert and so I deferred to Mahmoud Samy and Ahmed Maher, among other activists, and asked them to coordinate with each other" (Ghonim 2012: 144). Hence, the RYC's experienced activists were the principal co-ordinators of different groups on the ground and maintained order before the Egyptian protests. Ghonim did not take a role in mobilising and co-ordinating the public on the ground. When I asked Mona[7] who had been in the RYC and what they had done to organise the protests on the ground, she replied:

> We planned the marches. Where will the marches start? Who are the people that you will be responsible for? The thing is that you cannot just go out and protest. Only five or six people from the organising group can be in the same place at the same time. So, we divided amongst ourselves. How many people will be responsible for a specific spot? How will we release the statements? What will we write? So, our role was to go to some neighbourhoods before the January 25, talk to people and hand out brochures that said "we are having a protest on the 25th against the police".

The words of Mona reveal that the activists' past experiences were crucial in determining the protest performances of the Egyptian protests. Hence, as with Youth for Change (Al Shabab min Ahl Al Tagheer), whose members targeted the working-class Cairo community and visited public squares or parks in groups of four or five people (for full discussion of these protests please see Chapter 2), the Revolutionary Youth Coalition engaged with the public in different neighbourhoods (Azimi 2005). In particular, they succeeded in activating citizens who did not have access to the Internet. These experienced activists on the ground convinced the poor segments to join the protests by addressing to their problems. Thus, as in the workers movement in April 6, the urban and educated youth of Cairo contacted with the poor segment and asked to protest with them for bread, jobs and social security.

Mona's words also show that information about the protests was disseminated via conventional organisational tools such as brochures and statements. Along with these conventional tools, information was distributed via interpersonal communication. Tufekci and Wilson (2012), who surveyed 1050 Tahrir participants in January and March 2011, also found that people learned about the protests primarily through interpersonal communication using Facebook, phone contacts or face-to-face conversations. However, Mona's words reveal that face-to-face communication was particularly significant for the organisers in creating social solidarity. As Tarrow states, "people do not risk their skin or sacrifice their time to engage in contentious politics unless they have good reason to do so" (Tarrow 2011: ii). Direct contact helped organisers convince people that they had common purposes engaging in protests. For this aim, they were divided according to their social and cultural backgrounds. Each group would then focus on areas closed to their own culture and these divisions would facilitate communication between the organisers and public. According to Mona,[8] this offline method was the most effective mobilisation technique during the protests, which she summarised as follows:

> The Internet was only useful among us, the youth who were using the Internet. They always say that this is a Facebook revolution. That is not true at all! Maybe it is seen like that by the west. Online is useful but it is not enough to mobilise people. What is more important is going to the neighbourhoods that have problems with the police. For instance, I went to the Omrenea. This is an area where Christians have big problems and many had been previously killed here over the bombing of the church. When I went there and told them that they should go out on the 25th, they said they would go. They then mobilised the whole neighbourhood to join the protests.

Mona's account highlights the abstraction of the technological enthusiasm. When I conducted the interview with Mona, in 2014, she had already encountered several articles describing the Arab uprisings as "Facebook revolutions". In our interview, she felt it necessary to warn technological enthusiasts against the limited capacity of the Internet for mobilising protesters and highlighted the continuing importance of face-to-face communication in today's digitally supported movements. She also emphasised that going to areas with social and economic problems and personally making contact with the public were crucial techniques for assembling the masses. As was stated in this Chapter, social movement theorists have long emphasised the importance of shared grievances in creating collective action. Interest alone cannot trigger contention (Tarrow 2011: 11). Traditionally, it is believed that intellectual leaders emerge from different backgrounds and, given personality and experience, need to be identified with the oppressed and deprived groups (McCarthy and Zald 1977: 21). We can see that this hasn't changed in digitally supported movements; to develop a common interest, experienced activists on the ground acted as traditional leaders. They went to the areas with which they were identified and touched upon the existing grievances in these areas.

Another tactic used by the experienced activists, particularly in Mahalla, where worker protests between 2006 and 2009, was co-ordination through key figures in the villages. The Egyptian activist, Mamdouh,[9] recounts what he did for the organisation of the process:

> I had to make contact with every single village. I had 25 villages around Mahalla, which is like a small country. It is very difficult to go to every single place. So, you have to reach the key people, maybe a teacher or an engineer. They all had to come to Mahalla city during the revolution.

This interpersonal communication between key people shows that in the age of ICTs; opinion leaders are still key for information dissemination in authoritarian countries. With their theory of the two-step flow model, Katz and Lazarsfeld (1955: 25) argue that the response of an individual to a campaign depends on his social environment and the character of his interpersonal relations. Thereby, in order to activate a population, ideas first need to flow from the mass media to opinion leaders and from them to the less active sections of the population (Katz and Lazarsfeld 1955: 32). In El Mahalla, Mamdouh and his friends did not wait for the mass media to move from the news to opinion leaders, but personally went and spoke with them. One reason for this choice was that few media channels were reporting on upcoming protests. Secondly, opinion leaders in the villages had more networks than anyone else and so could affect the attitudes of the villagers more effectively than the mass media.

Experienced activists are separated from soft leaders in the section below, not because the soft leaders were inexperienced, but because not all experienced activists acted as soft leaders in the online sphere. For instance, Mona mentioned that she was not active as an administrator of a social media page on the Internet. In order to reach the masses on the Internet the experienced activists depended on the KKSFP administrators, they co-operated with them and benefitted from a large number of followers of that page. Mona[10] said:

Khaled Said was the main page. The other pages were getting the links from there and publishing them as well. The reason for this was, the KKSFP was the only anonymous page back at that time. The authorities knew the administrators of other pages. It had a good reach with thousands of followers.

Hence, although Wael left the co-ordination of protests on the ground to Ahmad Maher and Mahmoud Samy, he and the other Khaled Said admins played the main role in setting the scene for movement mobilisation in the online sphere. Mona mentioned that, along with the KKSFP, they co-operated with the ElBaradei Facebook page and Twitter to distribute specific posters for the Egyptian revolution to appeal to middle-class Internet users. They also performed unfamiliar practices such as distributing incorrect information about the protest sites on the Internet in order to baffle police officers by misdirecting them to non-protest sites during the 2011 Egyptian protests.[11]

Hence, in the Egyptian case, it was difficult to separate the action on the ground from the action on the Internet, the organising group successfully combined both online and offline mobilisation techniques to reach different segments of the society. The leaders in both the online and offline spheres divide the labour of mobilisation amongst each other. Each leader tries to ignite collective identity within the one particular segment of society on which she/he focuses. The experienced activists who mobilised the public on the ground, such as Mona, did not act as choreographers of protests in the online sphere; they left the communication and co-ordination of the online protests to the social media administrators. In the next section, I will focus on this second type of leader.

THE SOFT LEADERS

Similarly to what happened in Syria, a Facebook page, "Kullena Khaled Said" (We are all Khaled Said), sparked the rapid countdown to the Egyptian revolution in 2010, a year before the organising body on the ground started to meet. Wael Ghonim, a 29-year-old Google marketing executive, launched the Facebook page and dedicated it to Khaled Said, who had died in police custody in Alexandria in June 2010. The page contributed to motivating and uniting a middle-class segment of young Egyptians who normally do not have strong political identities

(Gerbaudo 2012: 55). My interviewees either heard about the 25 January 2011 protests from this page or had been amongst the organisers of the protests and in constant contact with the page admins. The success of the Facebook page in attracting the attention of the world and activating middle-class youth was linked to several important factors. Here, I focus upon experience, collective identity, discourse and the role of admins.

Experience

Unlike the Syrian Revolution Facebook page's administrators, this was not the first time that Wael Ghonim and Abdelrahman Mansour (the second administrator) had managed a page. Both had previously been admins of the ElBaradei Facebook page. The third administrator of the Khaled Said page, Sami,[12] who had helped Wael and Abdelrahman prior to the 25 January and would become the sole administrator of KKSFP when Wael went to jail and Abdelrahman to the army during the revolution, also returned to Egypt during the ElBaradei campaign. He joined with the hope of helping to bring about change. Thanks to their works on the ElBaradei campaign, the admins had already established a relationship with the journalists.

While organising the 2011 protests, Kullena Khaled Said admins did not need to learn online protest organisation, they already knew the journalists they should be in touch with. Accordingly, the administrators were very careful about hiding their identities on the KKSFP. My interviewee, Sami mentioned that he did not know that Abdelrahman was one of the admins in the first months:

> I met Abdelrahman in July 2010. He was administrating the page with Wael and I was in touch with the page through an anonymous account. Abdelrahman had this anonymous account called El Shaheed.[13] I was talking with them online and they were answering me, but Abdelrahman did not tell me he was one of the admins of the page.

The words of Sami reveal that the administrators were careful about hiding their activities even from those in their close circle. Working anonymously was not only crucial for the safety of the administrators and their families but also for the reputation of the page. Unlike the admins of the Syrian Revolution Facebook page, Wael and Abdelrahman knew well that the anonymity was important for the identification of the followers with the page. Abdelrahman[14] described the importance of anonymity for the administrators:

> If you stay anonymous, anyone can relate himself or herself to you despite your ideological differences.

This statement by Abdelrahman reveals that online, anonymity not only helps to avoid detection by the police but becomes a discursive opportunity for administrators. As Kang et al. (2013) mention, people might use the anonymity to create different personas online than they exhibit offline. The administrators of the Kullena Khaled Said page also used the opportunity that online communication offered to connect with their followers.

Collective Identity

Unlike the Syrian Revolution 2011 Facebook page, the KKSFP did not directly aim to mobilise people for a revolution. Wael Ghonim dedicated the Facebook page to a specific event, the murder of Khaled Mohamed Said. Said had been a 28-year-old entrepreneur from Alexandria. In June 2010, he was beaten to death in an Internet cafe by two police officers for possessing video footage of policemen sharing the spoils of a drug bust ("Khaled Said" 2012). The death of Said symbolised the brutality of Mubarak's police state, Wael Ghonim naming the account of the KKSFP "El Shaheed", which means martyr in Arabic, tried to create a common grievance amongst the supporters of the page. According to Sami,[15] the third admin of the page, what differentiated Khaled Said's case from the other victims of brutality was his middle-class background:

> At that time, there were so many activist pages that mobilised people and they had lots of audience, but we were lucky since there was big sympathy that grew over the Khaled Said case. He was a middle class person and the audience could easily identify with him.

As the members forged moral and emotional connections with the memory of Said, they contemplated the fact that this could also happen to them. For instance, after seeing the photo of Said's mangled face, Amro Ali, an Egyptian Ph.D. candidate, described his feelings:

> It was not just the manner of Khaled's death that disturbed me, but the
> deep reach of President Hosni Mubarak's repressive police state into a
> neighbourhood where I grew up and idealised as a beacon of harmony. Up
> until then, I naively thought that such things happened to other people,
> in the slums, Islamist strongholds, in prisons, on the news, Alexandria's
> rural outskirts, or any "other area". My area became that "other area".
> (Ali 2012)

The words of Amro Ali show how people can feel affiliated with individuals with whom they share space and community. What happens to
these individuals affects them more than others. Moreover, as Polletta
and Jasper (2001: 290) state, "prior ties motivated participation through
norms of obligation and reciprocity". The ties of Khaled with Alexandria
and the middle-class segment in Egypt pushed young people from the
middle class to like the page. These supporters created a snowballing
effect and also motivated other young people from different cities to
react to the Khaled Said case. Movements require solidarity to act col-
lectively and consistently (Tarrow 2011). By creating solidarity around
the figure of Khaled Said and with the reification of a "we", the administrators constructed a shared identity amongst the members of the page.
This identity rejected all sorts of torture cases and injustice that was
ongoing in Egypt.

Discourse

The language used on the page was also crucial to forming a collective
identity amongst the supporters of the page and the Khaled Said case.
Rather than adopting formal classical Arabic, Wael and Abdelrahman
purposefully chose a language closer to the heart of young Egyptians.
For this aim, they used the everyday language of young people, the colloquial Egyptian dialect on the page that would be understood much
more easily by his followers. In order to fortify the collective identity,
they avoided any kind of expressions that were not commonly used by
average Egyptians (Ghonim 2012: 61). This facilitated the process of
identification between the members and Said. They started to recognise
Khaled as one of them who shared the same youthfulness, middle-class
lifestyle, jokes, knowledge and emotions (Herrera 2014: 52). Ali (2012),
the Egyptian Ph.D. candidate, defined the attachment fostered through
the page as having created a myth that penetrated deeply into the blood

and soul. Yet, the identification of the page members with Khaled Said could not be seen in other such cases. Abdelrahman (cited in Herrera 2014), for instance, said that:

> We tried to shed light on other torture cases but, unfortunately, people responded to the cases from the middle class more. Khaled Said was from the middle class and many others were from poorer classes. They were tortured but no one talked about them. So, we made it clear that all human beings should have their dignities and we should defend anyone exposed to any kind of torture.

As explained above, Khaled Said's identity and his social and cultural status, were the main factors that caught the attention of the country's middle-class Internet-savvy youth. The discourse on the page aimed to direct this attention from the middle class towards a multitude of injustices prevalent in the society and to unify young people from different social segments. As the connective action theory of Bennett et al. (2014: 232) proposes, the Internet users "engage with issues largely on individual terms by finding common ground in easy-to-personalize action frames". In order to attract the attention of participants on the Internet, the leaders needed to touch upon the subjects with which the participants could identify themselves.

What particularly differentiated the KKSFP from other pages that were also dedicated to Khaled Said was its non-ideological discourse. The admins chose to embrace a non-ideological position to demonstrate that an organisation, political party or movement did not manage the page. Sami mentioned that Wael Ghonim's marketing experience in Google was crucial in this choice:

> Wael was very careful that the discourse was not going to be radical, it was not going to be ideological and it was not going to use activist language. We were not saying words against the police or Mubarak, like "Mubarak must go", until the 14th of January 2011.

Thus, adopting a non-ideological discourse helped the administrators to further identify themselves with the supporters of the page, who had hitherto been reluctant to interfere in political activities. The page had a frequently asked questions section in which Wael and Abdelrahman responded to questions under the pseudonym of El Shaheed—the

Martyr. Wael wrote in the first person, answering the questions as though he was Khaled Said. As social anthropologist Linda Herrera (2014: 53) mentions in her book "Revolution in the age of social media", Wael used this anonymity to create an aura of "every youth". He showed that the administrator shared the same non-ideological stand and common concerns about life in Egypt as the followers of the page. El Shaheed also mentioned that he did not belong to a political party, an organisation or a movement; he was just working with other volunteers to draw attention to police brutality and the "Emergency Law" in Egypt (Herrera 2014: 53). This non-ideological discourse helped embrace a larger number of people from different ideological backgrounds. From the youth of the Muslim Brotherhood, to leftist or liberal movements, people of different ideological positions supported the page and unified against police brutality in Egypt.

Launch Time of the Page

Unlike the Syrian Revolution Facebook page which was launched two weeks before the launch of protests, The KKSFP was launched on 8 June 2010 seven months before the Egyptian protests commenced. The time interval between the call for the page and the Egyptian protests in 2011 gave the administrators the chance to establish solid ties with the members. During this seven-month period, the administrators tried to strengthen the intensity of interaction amongst the members and to maintain the stability of membership. As the administrators were anonymous, they met with criticisms and suspicions during this period. Accusing and interrogative messages that questioned the ideology of the administrator and his/her affiliation appeared on the Kullena Khaled Said page too (Ghonim 2012: 91). However, Wael Ghonim's interactive and engaging dialogues with the members helped to sustain the page's popularity. Sami[16] ascribed this success to the amount of time they spent on the page with members, adding that:

> We were building trust relations with the audience through history. In this way, we could have an influence on them. If we created this Facebook page only two weeks before the protests, the audience would have thought "why would we like this page?" It was also about what the page was for. Our page was popular because people were sympathising with Khaled. These circumstances helped us. Each day we were developing the trust relations.

During this seven-month period, the administrators broke the barriers of fear by using different communication tactics. Wael Ghonim (2012: 68), for instance, mentions in his memoir that he asked the members of the page to photograph them holding up a paper that saying "Kullena Khaled Said". According to Wael Ghonim (2012: 68), these images posted on the page created an impact stronger than any words could have. The first members who agreed to publish their photos encouraged others to take the same action. Slowly, the members started to get to know each other and build trust relations. Hence, thanks to the speed of online communication, the administrators could access a large number of followers without expending immense effort, money and time on the ground.

The administrators also had time to create a new idioculture amongst the youth via online communication. We can refer to Fine to understand how the idioculture of a group is shaped: "the extent of a group's idioculture is connected to the length of time the group has been functioning, the social and psychological salience of the group to its members, the stability of the membership and the intensity of the interaction" (Fine 1995: 129). Thereby, the idioculture differentiated those inside the group from outside members. For instance, based on the suggestion of a member, the administrators agreed to organise silent stands. Sami described these silent stands as unique actions that differed from traditional activist protests:

> When the page called for action, it was not traditional activist action like a demonstration by which you confront the police and you usually get arrested afterwards. Instead, we called for a stand on the corniche of the Nile with enough distance between participants. Technically, it did not break the law. People did not chant and there were no expressive messages like reading the Koran or Bible. We were wearing black. Surprisingly, it was very successful and it attracted an audience that had never participated in activist movements prior to then. The stand was actually safe for them. We hadn't seen a lot of police except when the activists transformed these things into a march. When they gathered these people and took them for a march, then problems started to happen and that used to make Wael Ghonim very angry. He used to fight with the leadership of the activist groups such as April 6 and the Revolutionary Socialists.

Hence, in order to acquire the trust of the members and transfer online action to the offline realm, the administrators established a new idioculture that did not break the laws. This idioculture respected the non-ideological standpoints of its members. It also encouraged them to take action on the ground against the state without taking big risks. The attitude also prepared the members for further confrontations on the ground. Consequently, the administrators' activities in the online realm encouraged the middle-class segment to take to the streets and supplemented the offline protests.

Hybrid Leaders

Although Wael Ghonim was in Dubai and administrating the page from there, the two other administrators of the page, Abdelrahman and Sami, were in Cairo. They were measuring the "pulse" of the Egyptian public by attending the protests on the ground. Sami[17] stated that they constantly monitored what was going on in the Egyptian streets and reported them on the page.

> We were organising protests but also, whenever a page called for a protest, we were contacting the activist groups and I was going to these protests to take pictures. Afterwards, I created reports and sent them to our page.

Hence, although Wael was abroad, the administrators never lost touch with the activists on the ground. They also visited the families who had been subjected to police brutality and updated the public on them.

For instance, on 5 January 2011 a similar case to that of Khaled Said had occurred in Alexandria. Bilal Sayed from Alexandria was arrested by the police who then returned his dead body to his family the next day ("Update: Hundreds in Cairo and Alexandria" 2012). Sami, the third administrator, was doing an investigation on both the ground and in the online space. He stated that he got in touch with the family of Sayed and tried to find out more information about him. In this way, the administrators also worked as citizen journalists. They inscribed the grievance formed around the Khaled Said case in frames that identified other torture cases as well.

In addition, the administrators of the Khaled Said page did not wait for the activists to inform them about their plans. They were, for instance, actively present in the daily meet ups to plan National Police Day (25 January). Sami stated that, although he took part in the organisation of the protests on the ground, he also informed the admins about what had been said at these meetings. Hence, the administrators were very well informed about the aims and demands of the experienced activists who took part in the organisational process.

It is also important to note that some of the experienced activists who participated in the preparation of the protests on the ground had already held the soft leader role, which created the conditions for an assembly. They were also highly familiar with essential tactics for information dissemination on the Internet. In our interview, Esraa Abdel Fattah,[18] for instance, who was very well known by the public as the administrator of the April 6 Facebook page, stated that she spread the word about protests on her own social network accounts:

> I took the link of the Kullena Khaled Said page and wrote it everywhere. I wrote it on my profile, on my Twitter account, on the April 6 page and on the ElBaradei page. We used the photos of Khaled Said. These photos were the main reason that people joined the page. We opened discussions and tried to talk with people. I said, "that is Khaled Said; he is not a politician but a normal man. Participating in political life should not bring the policemen to come to your house to kill you. We will revolt against this. Khaled Said was not a member of a movement or party. He was a normal man like you". That was a very strong message that pushed people to join the page.

As Della Ratta and Augusto (2012) declare, technology provides the tools that made peer production and mass sharing possible. The activists who benefitted from the peer production and sharing culture of the Internet disseminated the images of Khaled Said amongst Internet users. These images were then reproduced in the guise of a meme and moved across space at implausible spread (Herrera 2014: 99). A meme could be defined as "an idea, behaviour, style or usage that spreads from person to person within a culture" (Herrera 2014: 116). In the case of Khaled Said, both anti-government ideas and behaviours diffused on the Internet. These memes even reached Internet users who were not members of the KKSFP. As Esraa mentioned, in order to encourage people it is crucial to approach them by referring to their problems. Being aware of this fact, the hybrid leaders contacted Egyptians online and offline and tried to create a common problem with the message, "what happened to Khaled can happen to us".

Another founder of the April 6 movement, Asmaa Mahfouz, was particularly active in setting the scene both in the online and offline realms.

Following the example of Mohamed Bouazizi, some Egyptians had set themselves on fire and their deaths had motivated some activists to take to the streets on the 18th of January 2011, Asmaa was one of these. Affected by the low level of attendance at the protests, Asmaa posted a video of herself on her Facebook account through which she encouraged people to attend the 25th of January protests. The video later appeared on YouTube.[19] It was also shared by the Khaled Said Facebook page and by mainstream newspapers such as *New York Times*, increasing its visibility (Wall and Zahed 2011: 1337). In his memoir, Ghonim (2012: 155) states that the video was crucial to breaking the barriers of fear of many Kullena Khaled Said page members. Unlike some brash Egypt bloggers with anonymous names, Asmaa shouted her support for the protests to a camera that was quite close to her face (Wall and Zahed 2011: 1337). Thus, she was completely visible in the video. As a young woman, her courage motivated many Egyptians and her image came to symbolise courageous Egyptian women.

Hence, the hybrid leaders often acted as "mobilisers", inspiring participants and also as invisible articulators. Some, such as Asmaa Mahfouz, used self-created videos in "vlog" style. This refers to videos that feature people looking into the camera with nothing else on the screen (Wall and Zahed 2011: 1335). The non-anonymity of Asmaa Mahfouz on social media did not prevent her appealing to different segments of society. Asmaa drew on gender identifications and appealed to male protesters to join the protests and protect the girls. According to society's embedded patriarchal ideas about honour, it was expected that an honourable man would step up to defend a woman who was in harm's way (Wall and Zahed 2011: 1339). The image and language of Asmaa, a normal but courageous Egyptian woman attending protests against the state by herself, galvanised public support. Hence, like the strategy adopted by Martin Luther King, Asmaa linked the aspirations of different Egyptian classes to those enshrined in Egyptian culture (Morris and Staggenborg 2002). She created a public self without relying on traditional media, as charismatic leaders and the leaders of professional social movements have to. She rather used the Internet and interpersonal communication on the ground to appeal to the cultural values of the public and mobilise them.

Today, by being the choreographers of protests in both online and offline spheres, hybrid leaders offer a rich array of mobilisation techniques. According to Tilly (2016: 40), if a movement strongly adopts familiar performances and also benefits from unusual practices, the movement's members come to prefer the flexible repertoire, which he calls "strong". In the case of the hybrid leaders, along with familiar performances, this group adopt new mobilisation techniques, which can be called a strong repertoire of contention.

The rapid diffusion of new communication technologies creates a pressing need to reconsider the complex and multifaceted forces that are reshaping the political communication environment (Chadwick 2013: 3). In Egyptian protests, the analysis demonstrates that three kinds of leaders shaped the political communication environment: the soft leaders, such as the admins of the social network pages, composed the first group; experienced activists and the members of political parties who organised the movement on the ground could be labelled the second group; finally, the hybrid leaders who were administrators or influential figures of the social networks but who also actively participated in the meet ups on the ground formed the third group. The hybrid leaders were effective in linking the soft leaders with the experienced activists as Kullena Khaled Said's administrators Abdelrahman and Sami did. All three of these groups were present in the Egyptian protests and played important roles in transforming contention into successful movement.

Conclusion

Tracing the organisation process of the 2011 Egyptian uprising showed that the Egyptian activists acquired distinctive protest experiences in the online and offline sphere during the ElBaradei campaign. Although the campaign itself and the silent standings did not translate into any success in the political realm, the experience and knowledge gained during this campaign and protests was crucial for the success of the Egyptian protesters. The ElBaradei campaign would help the activists of the 2011 Egyptian protests gain experience organising and preparing collective action on the ground and social media.

Second, the findings showed the significance of three types of leaders for the coherent organisation of a collective action. Acting as strategic decision-makers on the ground, the first group, experienced activists on the ground reached the poor segment and encouraged them to persist in high-risk activism. They effectively managed divisions of labour, set movements' goals and led the protests offline. Although some of them also operate on the Internet, they left the online communication and co-ordination of the protests to the soft and hybrid leaders.

Acting as e-fluentials and opinion leaders on the Internet, soft leaders provided emotion management amongst the middle class. Unlike the leaders of professional social movement organisations, as defined by McCarthy and Zald (1977), the "soft leaders" did not expose themselves on television to form the impression of widespread activity and grievance. Rather, they preferred to remain anonymous and operated behind the scenes on the Internet. The anonymity the Internet provided was important for two reasons—firstly, it was secure against an oppressive response by the state and, secondly, it attracted the support of large segments of society.

The third kind of leaders were the hybrid leaders who promoted collective action both on the ground and on the Internet. The difference from the soft leaders is that they themselves went to areas with problems and reported the issues from first-hand using ICTs. They also mobilised the public on the ground by distributing leaflets, brochures and face-to-face communication. Hence, the hybrid leaders blended together the pre-existing protesting styles of experienced activists with new media logics of soft leaders. They also became intermediaries between the soft leaders and experienced activists on the ground and facilitated the communication and co-ordination of the protests.

The repertoire of these leaders also changed. While soft leaders benefitted from ICTs as the new stitching mechanism, the experienced activists preferred traditional communication methods such as face-to-face communication or the distributing of leaflets and brochures prior to the Egyptian protests. They conveyed their messages online with the help of hybrid leaders who recombined supposedly old and new media logics and connected with different segments of society. Egypt was an important example of how these three leaders work in synergy during the mobilisation of protests.

Finally, the analysis showed that to involve people effectively in digitally supported protests action, creating a collective identity amongst the followers in the online and offline spheres was important. This chapter displayed that by dividing into groups according to their social and cultural backgrounds, experienced activists unified segment of society closed to their own culture around secular, non-ideological aims on the offline sphere. On the other hand, the tactics of the soft and hybrid leaders enable collective identity amongst the Facebook page followers, but that this process requires time and dedication. While creating their networks, the administrators of the KKSFP spent over a year establishing trust relations between them and Internet users. They secured this connection by creating a collective identity around the image of Khaled Said, who came from the middle class, as did many of the followers of the page. As Bennett et al. (2014) argue, individuals only engage in issues on the Internet on which they perceive common ground; although the admins tried to draw their attention to other torture cases that occurred in poor areas, they could not gain the support of the followers, as in the Khaled case. Unlike other social media pages that were also dedicated to Khaled Said, Wael utilised the language of this middle-class segment in his online posts and a non-ideological discourse. Appearing "authentic" and "one of us" are key determinants for being considered credible (Haslam et al. 2011). These discursive tactics thus helped the administrators to appear credible and unify their followers against many torture cases and injustices ongoing in Egypt. Moreover, the page administrators prepared their followers for possible collective action by organising safe protests in Cairo, referred to as silent standings. These protests could be labelled *connective* actions, as they were organised via the use of social media channels (Bennett and Segerberg 2012: 16). However, the arguments of third admin Sami indicated that the organisation was not carried out via "self-organised networks" on the Internet, but rather planned in detail through the administrators. Hence, this proved that leaders and collective identity were not just important for the organisation of past collective actions but also for that of digitally supported protests. These findings also displayed that today's digitally supported protests are not necessarily prototypical but, rather, comprised of actions with distinct characteristics and actors.

Notes

1. Interview with Mona (pseudonym), Egyptian activist, Cairo, 2014.
2. Interview with John Albert, Egyptian activist, Alexandria, 2014.
3. Interview with Esraa Abdel Fattah, Egyptian activist and blogger, Cairo, 2014.
4. Interview with Esraa Abdel Fattah, Egyptian activist and blogger, Cairo, 2014.
5. Interview with Esraa Abdel Fattah, Egyptian activist and blogger, Cairo, 2014.
6. Interview with John Albert, Egyptian activist, Alexandria, 2014.
7. Interview with Mona (pseudonym), Egyptian activist, Cairo, 2014.

8. Interview with Mona (pseudonym), Egyptian activist, Cairo, 2014.
9. Skype Interview with Mamdouh Salah Eldeen, Egyptian activist, 2014.
10. Interview with Mona (pseudonym), Egyptian activist, Cairo, 2014.
11. Interview with Mona (pseudonym), Egyptian activist, Cairo, 2014.
12. Interview with Sami (pseudonym), Egyptian activist, Cairo, 2014.
13. https://bit.ly/2FO9YkB.
14. Interview with Abdelrahman Mansour, the administrator of the Kullena Khaled Said page, London, 2015.
15. Interview with Sami (pseudonym), Egyptian activist, Cairo, 2014.
16. Interview with Sami (pseudonym), Egyptian activist, Cairo, 2014.
17. Interview with Sami (pseudonym), Egyptian activist, Cairo, 2014.
18. Interview with Esraa Abdel Fattah, Egyptian activist and blogger, Cairo, 2014 (refer to appendix).
19. https://www.youtube.com/watch?v=SgjIgMdsEuk.

REFERENCES

Abdelrahman, M. (2014). *Egypt's long revolution: Protest movements and uprisings.* New York, NY: Routledge.

Ali, A. (2012, June). Saeeds of revolution: De-Mythologizing Khaled Saeed. *Jadaliyya.* Retrieved from http://www.jadaliyya.com/pages/index/5845/saeeds-of-revolution_de-mythologizing-khaled-saeed.

Azimi, N. (2005, September 1). *Egypt's youth have had enough.* Open Democracy. Retrieved from https://www.opendemocracy.net/democracy-protest/enough_2794.jsp.

Bennett, W. L., & Segerberg, A. (2012). *The logic of connective action.* Cambridge, UK: Cambridge University Press.

Bennett, W. L., Segerberg, A., & Walker, S. (2014). Organization in the crowd: Peer production in large-scale networked protests. *Information, Communication & Society, 17*(2), 232–260. https://doi.org/10.1080/1369118X.2013.870379.

Castells, M. (2012). *Networks of outrage and hope: Social movements in the Internet age.* Cambridge, UK: Polity Press.

Chadwick, A. (2013). *The hybrid media system: Politics and power.* New York, NY: Oxford University Press.

Cole, J. (2014). *The new Arabs: How the millennial generation is changing the Middle East.* New York, NY: Simon & Schuster.

Della Porta, D., & Diani, M. (2006). *Social movements: An introduction.* Oxford, UK: Blackwell.

Della Ratta, D., & Valeriani, A. (2012). Remixing the spring!: Connective leadership and read-write practices in the 2011 Arab uprisings. *Cyber Orient, 6*(1). Retrieved from http://www.cyberorient.net/article.do?articleId=7763.

Elections in Egypt. (2010, November 23). Human Rights Watch. Retrieved from https://www.hrw.org/report/2010/11/23/elections-egypt/state-permanent-emergency-incompatible-free-and-fair-vote.

Fine, G. (1995). Public narration and group culture: Discerning discourse in social movements. In J. Hank (Ed.), *Social movements and culture* (pp. 127–143). London, UK: Routledge.

Gerbaudo, P. (2012). *Tweets and the streets: Social media and contemporary activism*. New York, NY: Pluto Press.

Ghonim, W. (2012). *Revolution 2.0*. Boston, MA: Houghton Mifflin Harcourt.

Haslam S. A., Reicher, S., & Platow. J. M. (2011). *The new psychology of leadership*. East Sussex, UK: Psychology Press.

Haulohner, A. (2010, February 20). Will ElBaradei run for president of Egypt? *Time*. Retrieved from http://content.time.com/time/world/article/0,8599,1966922,00.html.

Herrera, L. (2014). *Revolution in the age of social media*. London, UK: Verso.

Kang, R., Brown, S., & Kiesler, S. (2013, May 2). *Why do people seek anonymity on the Internet? Informing policy and design*. CHI'13, Paris, France. Retrieved from https://www.cs.cmu.edu/~kiesler/publications/2013/why-people-seek-anonymity-internet-policy-design.pdf.

Katz, E., & Lazarsfeld, P. F. (1955). *Personal influence: The part played by people in the flow of mass communications*. New York, NY: The Free Press.

Khaled Said: The face that launched a revolution. (2012, June 6). *Ahram Online*. Retrieved from https://bit.ly/30uwE0G.

Kirkpatrick, D. D. (2012, January 14). Nobel laureate drops bid for presidency of Egypt. *The New York Times*. Retrieved from http://www.nytimes.com/2012/01/15/world/middleeast/mohamed-elbaradei-pulls-out-of-egypts-presidential-race.html?_r=0.

McCarthy, J. D., & Zald, M. N. (1977). Resource mobilization and social movements: A partial theory. *American Journal of Sociology, 82*(6), 1212–1241.

Morris, A., & Staggenborg, S. (2002, November). *Leadership in social movements*. Retrieved from https://www.sociology.northwestern.edu/documents/faculty-docs/faculty-research-article/Morris-Leadership.pdf.

Oberschall, A. (1973). *Social conflict and social movements*. Essex, UK: Pearson Education.

Poell, T., Rider, B., Woltering, R., Liesbeth, Z., & Abdulla, R. (2015). Protest leadership in the age of social media. *Information, Communication and Society, 19*(7), 994–1014.

Polletta, F., & Jasper, J. (2001). Collective identity and social movements. *Annual Review of Sociology, 27*, 283–305. https://doi.org/10.1146/annurev.soc.27.1.283.

Preston, J. (2011, February 5). Movement began with outrage and a Facebook page that gave it an outlet. *The New York Times*. Retrieved from http://www.nytimes.com/2011/02/06/world/middleeast/06face.html.

Shehata, D. (2011, May/June). The fall of the pharaoh: How Hosni Mubarak's reign came to an end. *Foreign Affairs*. Retrieved from https://www.foreignaffairs.com/articles/north-africa/2011-04-14/fall-pharaoh.

Siperco, I. (2010, July 17). Egypt's undeclared change candidate: The symbolic appeal of Mohamed ElBaradei. *Middle East Policy Council*. Retrieved from http://www.mepc.org/articles-commentary/egypts-undeclared-change-candidate-symbolic-appeal-mohamed-elbaradei.

Tarrow, S. (2011). *Power in movement: Social movements and contentious politics*. New York, NY: Cambridge University Press.

Tufekci, Z., & Wilson, C. (2012). Social media and the decision to participate in political protest: Observations from Tahrir Square. *Journal of Communication, 62*(2), 363–379. https://doi.org/10.1111/j.1460-2466.2012.01629.x.

Wall, M., & Zahed, S. (2011). "I'll be waiting for you guys": A YouTube call to action in the Egyptian revolution. *International Journal of Communication, 5*, 1333–1343.

CHAPTER 4

Resource Mobilisation Strategies Retested

From the Arab uprisings in 2011 to the Romanian protests in 2018, we have seen that the collective efforts of crowds who mobilised via ICTs often have practical limits. The digitally supported volunteer networks struggle, as they have sidestepped some of the traditional tasks of organising. Online activists often came together in the offline sphere for a short period of time, expressing the need for change, but this remained an abstract signifier in real life (Adi et al. 2018). Unlike most digitally supported protests, Egyptians exhibited an incredible capacity to mobilise, co-ordinate and sustain crowds over the 18 days of the protests in 2011 that ended with the overthrow of Mubarak. The organisers struggled to achieve their primary aims, such as freedom and social justice after the 2011 revolution as well. Most of them also indicated that their revolution failed in fulfilling their demands (Eskandar 2015). I discuss the reasons behind this in the conclusion section, and yet despite social change that never materialised in the country, the organisation and mobilisation of the protests during the 18 days were a good example of how ICTs can be effectively employed to create a solid, sustainable action against the state's activities and force a political change in the country. In Chapter 3, I explored the role of different leaders in the organisation of the Egyptian protests. The purpose of this chapter is to better understand how the protesters mobilised, co-ordinated and sustained the protests over these 18 days. Tracing the different stages of the protests and the response of the state to these is important for

93

B. Aslan Ozgul, *Leading Protests in the Digital Age*,
Palgrave Studies in Young People and Politics,
https://doi.org/10.1007/978-3-030-25450-6_4

identifying the key actors, their efficacy in using ICTs and the reason for this technological capacity.

Most research into the 2011 Egyptian protests and the role of ICTs either analyses how activists used social media during the protests or explores the impact of ICTs on Egypt's social and cultural environments over the long term (Hofheinz 2011; Howard and Hussain 2013; Herrera 2014). Only a few researchers, such as Eltantawy and Wiest (2011) and Tufekci (2017), have reversed the question and analysed the conditions that affected the use of ICTs in 2011. Analysing the Egyptian and Tunisian protests, Eltantawy and Wiest (2011) claim that in analysing today's movements, resource mobilisation theory should not be discounted: it places emphasis on the social, historical and political contexts of collective action. Resource mobilisation theorists such as McCarthy and Zald (1977) and Tilly (2006) emphasise the importance of resources, such as time, money, organisational skills and certain social or political opportunities, for the emergence and success of social movements (see introduction for more information on resource mobilisation theory).

By analysing the actors, their repertoires and the conditions under which they operate, I also aim to test the relevance of resource mobilisation theory in the context of Egypt. Eltantawy and Wiest (2011) report that the rise of social media as a novel resource in today's movements has enabled the rapid receipt and dissemination of information, helped to build and strengthen ties amongst activists and connected the protesters to the outside world (Eltantawy and Wiest 2011: 1218). Thus, the experience and knowledge gained from past protests proved helpful during the Egyptian protests. In line with the argument of these researchers, the first finding of this chapter shows that the past protest experience of the activists enabled them to effectively use ICTs in both the online and offline spheres. Based on their past protest skills, protesters utilised face-to-face communication to mobilise the poor segment of the society. ICTs were rather used to mobilise middle-class Internet users, buffer police officers, co-ordinate different groups and internationalise the movement. The findings of this research build on the current literature by showing that the effectiveness of ICTs not only depends on the past experience of the actors involved but also on the availability of organisations that provide facilities for the movement. The civil society in Egypt, along with political institutions, NGOs and human rights organisations, shaped the way in which ICTs were used in the Egyptian protests. Thanks to this organisational support, Egyptians also managed to create a non-sovereign sphere in Tahrir in which different ideological groups could coexist, pray and protest together. These findings are important because they will show that resources such as past protest experience and the networks of the actors, as well as their organisational facilities, are still important components for the sustainability of today's movements.

Mobilisation Tactics Under Three Styles of Leadership

The mobilisation of the 25 January protests was the most important part of the Egyptian revolution, bringing energy, support and dynamism to collective action. The movement was centralised around the capital city of Cairo but in fact, experienced Egyptian activists returned to their hometowns on 25 January, believing that they would be more active in their city during the protests. One of these was Sherif Azer,[1] who has worked as a human right activist in Cairo since 2003. His duty was to monitor the protests that took place for various reasons, such as freedom of expression, labour and journalist rights in Cairo. Sherif described the atmosphere in his town, Asyut, on 25 January:

> I thought something would happen in my hometown, so I wanted to be there. But not much was going on, so I took the train and came back to Cairo. By the time I arrived there, the police had already employed violence. My friends had already been beaten up. We took them to the hospital.

As Sherif claims, his town of Asyut, which is home to the country's large concentration of Copts,[2] witnessed the protests of only around 100 activists on the 26th (Protests in Egypt 2011). Similarly to Sherif, Abderrahman Ayyash[3] was in his hometown, Mansoura, on 25 January. He was a former member of the Muslim Brotherhood in 2011, and an influential blogger. Since 2006, he has spent most of his time on the Internet, where he observed that the digitally organised protests were ineffective in bringing Internet users to the Egyptian streets. Therefore, he did not expect big crowds on the streets on 25 January. Contrary to his expectations, a big crowd did gather in Mansoura on that day. During our interview in 2013, Ayyash described the scene:

> On the first day, there were 5,000 people in Mansoura, although we had been expecting to only be around 200. It was amazing for us. In Cairo, I was talking with some friends from the April 6 Movement. They told me that they had started with 70,000 people in Cairo. We thought that this had been the only successful event organised via Facebook till that point.

The words of Sherif and Ayyash demonstrate that, despite the mobilisation in some cities, such as Mansoura and Asyut, the central point of the Egyptian protests was Cairo. The careful mobilisation plans of the soft leaders, hybrid media activists and experienced activists on the ground concentrated the efforts of different protest groups in Cairo and created the biggest uprising in Egyptian history in the capital city. The plans of the organising group had been to commence the protests in different lower- and middle-working-class parts of Cairo. Constant movement of the protest groups was also an essential feature of this protest (interview with Esraa Abdel Fattah, 2011). The marches from different protest sites accelerated the protests and brought thousands onto the streets. People would join the protests when the protesters would come to their streets, making them and their demands central to these (Gunning and Baron 2013: 248). This method would also be used to split police forces. My interview with Mona[4] (pseudonym), who was in the organising team, exhibits that the protesters prepared themselves for big clashes with the police and for severe medical conditions they could experience in jail:

> We started to use the word "revolution" after Tunisia, but we were not really convinced that it would be one. It was only two or three hours before the protests when we discussed over the phone that we were going to bring people to Tahrir. Also, two nights before, I remember we gave all our medical records to lawyers because some of us, like me, have chronicle illnesses. We thought, on 25 January, we were all going to jail, so we listed the names of medications we needed.

Mona's words show that the Tunisian protests deeply affected the public and created an opportunity for the organisers that they had not imagined. As the organising group knew how to stage a demonstration on the basis of their past experience, they prepared themselves for several possible outcomes. Like the French people in 1789, who made collective claims by drawing on the available repertoire of contention, the Egyptian activists referred to their past repertoire and ensured that they would have medicine and lawyers in case they were arrested.

Moreover, in order to minimise confrontation with the police, the organisers hid information regarding where to assemble until the last minute and the messages were then distributed via mobile phones, text messages, social media or emails (Gunning and Baron 2013: 170). These types of mobilisation, in which groups of protesters spring into action from different places within a seemingly random crowd and then work in co-ordination via ICTs, are called flash mob tactics. The protesters had previously used these in the "Global Justice of Movement" in Egypt and had seen that it brought strength to a movement and gave courage to protesters. Influenced by these tactics, the groups remained dispersed and mobile communication then led them to converge on specific locations, in this case Tahrir Square. Mona[5] described this process:

> We announced the protests on the Khaled Said page and on our page [the ElBaradei Facebook page]. We said that we were going to work together on 25 January and that we would go out from certain areas. The names of some areas that we announced were true and some were not. For example, we said on the Internet and TV that there would be protests at Gameat El Dowal. On the 25th, I went out very early and there was a policeman at every corner of Gameat El Dowal. They thought that we would start from there, but we actually started from Nahia, which is a very crowded market area. They did not think that we could start from somewhere like that. That's why we had millions with us…

Mona's statement indicates that by drawing on their past protest experiences, the leaders knew that the police would monitor the Internet and mainstream media. In 2011, the protesters turned this disadvantage into an advantage and used ICTs to lead the police to false starting points. The exact information about the protest sites was distributed by hand, via leaflets and brochures. On the leaflets was written: "please distribute through e-mail printing and photocopies only. Twitter and Facebook are being monitored" (Gunning and Baron 2013: 284). The leaders clearly knew when to benefit from and when to avoid the new media channels. The mobilisation tactics were crucial information, and probably the last thing that protesters would have liked to share with the police.

During our interview, Mona also explained why specifically they selected a poor area such as Nahia as the starting point of the protests. Nahia was chosen since its residents had previously been subjected to police violence and also lacked basic services. This made them susceptible to social and economic grievances. Moreover, as a crowded area from where a large number of people could be mobilised, Nahia was the ideal starting point for the organising group. However, the face-to-face contacts of the organisers with the poor residents of Nahia showed that the residents would only go out if everyone else did. To overcome this problem, the organisers chose small streets of Nahia as starting points. Such small streets would make small numbers look larger and encouraged local residents to join the movement (Gunning and Baron 2013: 171).

Starting from small streets, though, also carried risks. In Alexandria, for instance, the police managed to place a spy in the organising group. John,[6] who was one of the organisers, explained the situation:

> On 26 January, we started the demonstrations in a very local area. We chose a narrow street because we thought that police cars could not enter it. On 25 January, this strategy worked but, on the 26th, the police knew where we were going to start the protests. They brought some soldiers and surrounded us. They then took us all out and arrested us.

Hence, while the small streets were effective for bringing large numbers of people out, they were also unsafe as the police could easily surround protest groups and arrest their members.

One crucial aspect of the Egyptian protests that has not been much explored by scholars is that there were a dialectical relationship and approach to co-ordination between the organisers and established political groups in the Egyptian protests. For instance, although Muslim Brotherhood leaders did not openly support the mobilisation process on 25 January, the protesters could actually benefit from these established political groupings. In an interview in 2012, Masud[7] (pseudonym), who was then editor of Ikhwan web (the Muslim Brotherhood's official page), said:

> We knew that the Muslim Brotherhood would not take a swift decision very easily and become involved in the protests because they had to take this up the hierarchy and get approval and this would take time. There were 55 people at Ikhwan web and I told them, "You should go and mobilise people to go to Tahrir. This may be our last battle. Forget about being affiliated with the Brotherhood. Just go and mobilise people to protest in Tahrir". We had around eight secret headquarters in different places. We allowed young people to use these secret places for mobilisation. They came to these places, spent the night and organised the protests. We provided them with laptops and cameras.

The words of Masud prove that the Muslim Brotherhood's leaders were reluctant to give an immediate response to the calls for protest. However, thanks to the youth and certain elites inside the organisation, the protesters benefited from the Muslim Brotherhood's resources. My young Egyptian interviewees, such as Ayyash, who was a member of the Muslim Brotherhood prior to 2011, also confirmed the words of Masud.

Resource mobilisation theories often emphasise the role of reform professionals such as bureaucrats and the members of charities and religious organisations in the movements (Della Porta and Diani 2006: 213). As the most established political force in Egypt with a long-running history, the Muslim Brotherhood had been an important element of Egyptian politics. Although it was officially outlawed, this religious organisation had been able to participate in political life through some of its members campaigning as independents. The Muslim Brotherhood also created strong social networks through the building of hospitals, pharmacies and schools in rural areas (Kuebler and Allagui 2011). While the resources of the religious organisation fortified the logistical and technological infrastructure of the Egyptian protests, its networks enlarged the level of participation in these.

In particular, in Alexandria the Muslim Brotherhood (Ikhwan) members were effective in widening the mobilisation of the protests. Egyptian activist John Albert[8] claimed that:

> The ElBaradei campaign only had 10 or 15 people in Alexandria, and there were also people from the April 6 movement but, actually, we used to work with Ikhwan. Ikhwan was really effective. They possessed a huge number of people. They were asking us how many people were going to the protests. We would say, for example, 15, and they would send 15 people as well. We would respond, "You must have more people". And they said, "For security reasons, our number cannot be more than yours". I had not known any of them before, but when I joined the ElBaradei campaign, I met with these people and we really were not different. We wanted the same things.

John's statement revealed that the Egyptian activists not only created virtual communities but they also started to get to know each other and built networks on the ground before the 2011 protests. Conforming to the argument of Cole (2014: 149), who also explored the 2011 Egyptian protests and emphasised the role of personal networks in the mobilisation process, the interviewees for this research mentioned that the networks of different groups such as the April 6, ElBaradei and youth Muslim Brotherhood were more effective than the Internet in getting people to assemble.

On the ground, however, what made the difference was the experienced activists, who played crucial roles during the mobilisation of the Egyptian protests. The young members of Kefaya, April 6, the workers' movements and the Muslim Brotherhood, they all had been ready on the street since the beginning of the protests. For example, when I asked Ahmed,[9] a prominent researcher and ex-member of Muslim Brotherhood, why he joined the 25 January protests, he claimed that the experienced activists participated in the protests without much thought. He said:

> Let me be honest, I was sceptical about the outcome of this particular protest. I thought it would be a big demonstration but never a revolution. Thus, why did I join? I know that it is a very simple question but actually, I never thought about it because I was so used to do it. When there is a demonstration, I join it.

The words of Ahmed clearly reveal the inclination of the experienced activists towards the protests. Ahmed, who has been an active participant in protests since the 2000s, did not sense the same uncertainty that a first-time protester would experience. Although those like Ahmed were sceptical about the possible outcome of the protests, they were not afraid to be part of them. Their support also increased the number of participants. As social movement theorists, Della Porta and Diani (2006: 113) state, people are more likely to join a movement if they have links with others who are highly sensitive to particular causes. The experienced activists who were deeply sensitive to issues concerning the state fostered the participation of others by using their networks. A prominent Egyptian blogger,[10] for instance, told me:

> I was not politically active before the 25th of January. I only attended to a couple of rallies of ElBaradei and went to the demonstrations that were organised by the judges here in 2006 and 2007. But, I had many active friends who joined previous protests. I called them on 24 January and the next day I met with them to join in the protests.

The blogger's statement confirms that experienced activists would often guide action and encourage deeper engagement with the protests. Surely, some Egyptians joined the protests without being part of a group. Using the new capacities offered by ICTs, such as being informed about the protests and amassing one's network online, another Egyptian interviewee[11] of mine described how they tried to individually mobilise people:

> I was at the work on the morning of the 25th of January. Then, I went to Tahrir square. We were about 50 people. Some of them were my friends and some were from the community. We did not go directly to Tahrir. We started in Ramses street, which was two buildings away from Tahrir. At that time, the number of people in Tahrir square was not too great. We started to tell people to go to the square. By midday, thousands were running to the square.

The individual efforts of my interviewee, who had been working for an Egyptian media channel since 2011, and her friends are good examples of initiatives taken by protesters during the digitally supported movements. Individual protesters who were hungry for a change would go out based on information in the mainstream mass media or on the Internet and would also become advocates on behalf of the movement. However, for security reasons, being part of a protest group led by experienced activists and following protesters' social media accounts might be particularly important. For example, an activist from Alexandria, Mohammed,[12] joined the protests by himself and stated that:

> On 26 January, there were no demonstrations in the whole of Alexandria. I was walking down the street myself and looking for protesters. Then, the police took me to the station. The activists had actually agreed that they wouldn't organise any protests on the 26th and 27th and decided to save their strength for 28 January, which was a Friday. I did not know this. I stayed in jail until Saturday night.

The words of Mohammed indicate that the protesters who were not part of any protest groups and not engaged with social media accounts such as Kullena Khaled Said page were more likely to face police repression.

First Reactions of the State

Since information about the protests was in both the mainstream and online independent media, the Egyptian state was well aware of the possibility of conflict arising on 25 January. The police quickly intervened in the protests on that day, firing tear gas and using water cannons against the protesters. However, the size of the protests was beyond what the police had anticipated and so they were largely ineffective, particularly against the groups that had mobilised from secret starting points, such as Nahia and Boulaq Ad-Dakrour. As the police were unaware of these gatherings, Nahia and Boulaq Ad-Dakrour residents could carry out their march and occupied Tahrir until midnight before the police attacked them with tear gas and rubber ballets (Levinson and Coker 2011).

Along with Cairo, demonstrations also broke out in Alexandria, Suez, Ismailia, El Mahalla El-Kubra, Asyut and other towns on 25 January (Cole 2014: 148). As the protests spread in the peripheries, the Interior Minister issued a statement that blamed the Muslim Brotherhood for triggering the unrest (Gunning and Baron 2013). As a former Muslim Brotherhood member, Ayyash[13] was in the protests in Mansoura. He described the attitude of police when he was arrested:

> I was the last person that entered the state security building in Mansoura. They wanted to know if the Muslim Brotherhood was on the street or not. An officer came to us. We were 40 people strong and he was just looking at the names. You know, Ayyash is the third name on my ID. When he read my first and second names, Abdelrahman Mahmoud, he just raised his eyes and then he saw Ayyash there and shouted "AYYASH!!" He was happy. He thought that, if I was arrested, the MB was there and so he could say that the MB was arranging the protests. I said, "No, I am not with them anymore" but he did not believe me!

Claiming that MB was leading the protests, the government aimed to break the unity of the protesters and disdained the seculars who would not support a protest activated under the lead of MB. The discourse adopted by the state was also used in the state-run media that provided it with privileged access to promote this discourse. For instance, Osman and Samei (2012: 20), who performed an analysis of the state-run newspaper, *Al Ahram*, mentioned that it accused the Muslim Brotherhood of seeking to turn people against the police. As Miskimmon et al. (2013) claim, how states form and project a narrative in a communication environment help us explain their strategies in national and international politics. By controlling the narrative of state media, the Egyptian state tried to shape both Egyptian and international public interpretation of the protests and enforce the view that this was an uprising with an Islamist agenda.

When the national media in Egypt is analysed, two different narratives during the protests can be observed. On the one hand, there is state-run media, such as *Al Ahram*, that ignored the online news and informed on the protesters through a small report. On the other, we see that private press, such as *Al Shuruq*, provided extensive coverage of the protests and encouraged other protesters to take the street (Osman and Samei 2012: 20). Hence, despite the state-run media that tried to prevent the dissemination of the protests on a national scale, private newspapers were reporting the news from online networks and highlighting growing popular dissatisfaction with the regime.

The integration of foreign journalists into the protest groups also gave the protesters an important advantage against the state in the framing of the protests. Despite the fact that some of the foreign journalists were arrested and harassed by the state, the journalists were mostly able to report from the ground and covered the news during and after the revolution (Khamis et al. 2012: 20).

In terms of the Internet, since its introduction in Egypt, the Internet had actually been closely monitored by the Egyptian state. Although the government had opened up the Internet service provider (ISP) sector to private competition in 2002, the largest ISP of Egypt was the Tdata company and the company was owned and operated by the state. Thanks to these arrangements, the state could easily determine the websites that would be made available to public users. Moreover, in 2002, a special unit was established within the police department and specifically charged with monitoring the Internet (Ezzat 2014). The Egyptian journalist Wael Eskandar[14] also stated that he had worked for a company called Giza Systems in 2009 that was commissioned to install special software for intelligence services that looked into the Internet packages of citizens.[15] These monitoring attempts were crucial to determining the identity of bloggers who criticised the Egyptian authorities online. Many bloggers who dedicated themselves to uncovering government abuse faced unjust treatment at the hands of the Egyptian security officers. Prominent bloggers such as Wael Abbas or Abdel Kareem could, in turn, be cited as one of these arrested bloggers.

With the start of the 2011 Egyptian protests, the Egyptian state acted more rapidly on the Internet. In my interview with Mona,[16] she indicated that the state first started to cut mobile phone connections and Twitter access in Tahrir Square in order to isolate the area. It was also quick to initiate a censorship policy towards the websites of the newspapers that supported the protests. The Al-Dostour website, a liberal opposition news website, and Al-Badeel, a leftist news website, were amongst those censored websites. Several websites of the Muslim Brotherhood also became inaccessible on 25 January ("Egyptians report poor communication" 2011). Thus, the state tried to regain its privilege to shape public discourse by restricting opposing voices on the Internet.

Despite these restrictions, the state lost control over news coverage to social media on 26 and 27 January. My interviewees stated that proxy servers were used during the first two days to access Facebook and Twitter. Although not many Egyptians were familiar with this method, those who were could use proxies to transmit news of the protests to the mass media.

As argued by Felix Berenskoetter (2012: 19), creating a hegemonic master narrative and defining the limits of what was known and what was deemed possible was an important source of power during the protests. The Egyptian state progressively lost this power, as the protesters were capable of challenging and replacing the state's master narrative through their interactions in Egypt's hybrid media environment. Consequently, the state decided to regain the privilege to shape the discourse and communication by limiting access to social media platforms (Freedom House 2012). The connection problems with Twitter first started on 27 January and protesters interpreted it as a test before the shutdown of the Internet took place. In line with these expectations, the state shut down the Internet completely on the night of 27 January, just two days after the start of the protests. According to Freedom House reports (2012: 4), apart from the Noor Internet service provider that assisted Egypt's cabinet, public banks and the country's airways, the state shut down almost all of its "Border Gateway Protocol Routes", which detached the country from the global network. Cutting off the Internet was a major blow to the co-ordination strategies of the protesters as well.

CO-ORDINATION TACTICS OF PROTESTERS

Between 25 and 27 January, co-ordination between different protest groups was successfully conducted via mobile phones, Google maps and social media accounts. According to a survey conducted by researchers Wilson and Dunn of digital media users, the use of Facebook was ahead of Twitter and blogs (Wilson and Dunn 2011: 1252).

However, the interviewees for this research indicated that the protesters applied different digital media platforms to protest activities for different purposes. Ahmed Samih,[17] who heads the Egyptian NGO, the Andalus Institute for Tolerance, described the role of social media in the Egyptian protests: "Every direct communication was on Facebook. Facebook was like our political party and Twitter was our walkie-talkie". By political party, Ahmed indeed refers to Facebook groups that collected popular grievances and informed the media to get further support for the protesters (Khamis and Vaughn 2012: 12). Just like political parties, Facebook groups operated to achieve a strategic goal and inform the media and public about this. On the other hand, Twitter was used for the co-ordination of the protests at the protest sites. It was more useful for instantly updating the public about the continuous group of events throughout the crisis. During our interview, the human right activist Sherif Azer,[18] for instance, stated that:

> When I was at home, I could follow what was happening via Facebook but, once I was on the street, I don't think I used Facebook much. Only when Mubarak left, I updated my Facebook status and wrote "We did it!"

Papacharissi and Oliveira (2012), who analysed 1.5 million tweets with the #Egypt hashtag utilising Latin characters also found that Twitter's difference from other digital media channels was its instantaneity. With the help of this new media, activists recorded events and reported on them promptly. After all, finding the Facebook group unit to disseminate the information took a longer time than writing the hashtag of the protest on Twitter. These findings support the argument of Wilson and Dunn (2011: 1254) who state that, despite a low number of users, some digital media are inherently and noticeably well suited to use at protests compared to others. In the Egyptian case, Twitter was the digital media platform that was particularly apt for the protesters' co-ordination purposes.

Apart from face-to-face communication, mobile phones and SMS were the most used communication tools during the protests (Wilson and Dunn 2011). Consequently, when the Internet and mobile phone services were shut down on 28 January, the co-ordination process became much more challenging for the protesters. Sheriff Alaa,[19] for instance, described the chaos they endured in the absence of phone services:

> On 28 January, I went to the protests from home with my sister and a friend. We could not go to Tahrir because it was blocked. I went to a place that was close by known as old Cairo. It was not really my neighbourhood. When police started to use gas, my sister and I just stayed in the same place and kept calling to each other. I could not reach her with my mobile phone due to the cut and I did not go anywhere without her. When the tear gas went away, we realised that there were only a few people left in the square and we were all surrounded by the police.

Sherif's words underline the significance of mobile phones for increasing the safety of the protesters at the protest sites. In the absence of ICTs, the activists lost their connections with their groups and ended up in different areas with different people. The human right activist Sherif Azer,[20] for instance, described the co-ordination process as follows:

> It is completely random. We all went to Tahrir first. You would usually start protesting with the group that you met in the morning. You decide on your group for the day. When the tear gas comes, you can only find 2-3 people from your group and the rest gotten lost. Then, you find another group with completely different people.

Hence, by shutting down mobile phone connections and the Internet, the state significantly impeded co-ordinated action. In addition to the co-ordination difficulties, the protesters struggled to spread information regarding protesters who had been injured or killed. Consequently, the provision of medical and legal aid for the protesters was unmanageable.[21]

Nonetheless, these co-ordination difficulties did not disperse action but rather motivated the public to take to the streets. Scholars who analysed the media activities of the Egyptian state during the 2011 protests, such as Gerbaudo (2012), often claim that one of the biggest mistakes the state made was the Internet shutdown. The interviewees for this book also subscribed to this argument and said that the unplugging of the Internet connection galvanised the people on the streets. The shutdown particularly unified those who were sitting in front of their monitors, as well as those on the streets. A poll conducted by the International Republican Institute (IRI) also demonstrated that the participation level in the protests was particularly high on 28 January in comparison with the other days. According to this poll, 40% of the people attended the rallies for the first time on 28 January. This level of participation was even higher than that on the first day of the protests.

In explaining the high attendance level, my interviewees mentioned that while their friends and relatives were at the protests, they could not wait at home without any information. They therefore joined in the protests to prevent any possible violence that could break out between the protesters and police. The journalist Wael Eskandar[22] said:

> On 27 January, they shut down Twitter. Then, we started expecting the shutdown of other mobile phone services. I remember an engineer that I really respect saying, "I went to Tahrir because, when they cut down the communication, I thought they were treating people like animals".

Thanks to thousands who think in a similar manner to this engineer, the Egyptian state's communication cut tactics would backfire. The communication cut rather enlarged participation in the protests and tens of thousands of protesters took to the streets.

The careful organisation of the 28 January protests was another important factor that broke the barriers of the state and helped in the co-ordination of the movement in the offline sphere. As a member of the Revolutionary Youth Coalition, Mona[23] mentioned that when Twitter was shut down on 27 January, protestors started to expect that the shutdown of all mobile phone services would follow. In order to stay in touch with the others, they decided to collect the landline numbers of those in their close circles. The time and place of the protests were also determined in advance. The organising group decided the roads for the marches and then used Google Maps to update the people on them. Trained by NGOs, some of the activists from the April 6 movement learned how to use mapping tools, such as Google Maps, to document protests and determine demonstrations (Khamis and Vaughn 2012: 165). Esraa Abdel Fattah,[24] for instance, mentioned that:

> This was the first time (during the 25 January protests) we used Google
> Maps to find out how protestors need to walk until everyone meets in
> Tahrir at the same time. There was someone called Bassem Fathy. He was
> responsible for Google Maps. We also used this application on 28 January.
> We started to find the mosques, how people could walk from the mosques
> to Tahrir, which roads they needed to take.

Basem Fathy,[25] who was mentioned in this statement by Esraa, was previously a member of the April 6 movement and the ElBaradei campaign.
His experiences with these networks would help him use new means of
action within the Egyptian repertoire. He would use Google Maps, for
example, to determine the roads and took the print of these maps prior
to 27 January by taking a possible Internet blackout into account. In the
absence of the Internet, the mobilisation and co-ordination of the 28
January protests, thus, occurred smoothly thanks to the experience and
networks of the activists. As mentioned by Tilly (2006: 55), significant
changes in repertoire might occur in countries in parallel to the development of technology. Yet, no actors abruptly shift from one repertoire to
another, as it is a continuing process that follows innovation and modulation. The experience of the Egyptian protesters assisted them in using
the technology in the most effective ways and helped foster a continuous
process of innovation and modulation.

The words of Esraa also show that, after the first day of the protests,
the organising group concentrated their efforts on preparing the 28
January protests. These were important since the 28 January 2011 was
a Friday and these are holidays in Egypt. As Friday prayers are the largest
weekly gatherings of Egyptians from all classes, it was easier to reach the
different segments of society at these prayers (Ghonim 2012: 156). My
Egyptian interviewee, Mamdouh Salah Eldeen Amer,[26] who was one of
the protest organisers in Mahalla, said that:

> In the mosques, you can talk with people that you cannot reach normally.
> I think it was difficult to activate people because they were afraid. We
> were against the whole system; not against only one organisation, like the
> police. Being afraid is a natural feeling. We were also afraid but we had to
> do this. So, the mosques were important for reaching people.

Here, Mamdouh points to an important problem that social movement
scholar Melucci (1996: 295) also defines as one of two key problems

that a movement's leadership should address in order to widen participation: The first is the reduction of risk and the second the maximisation of advantage. In the absence of ICTs, the mosques helped the organisers reduce the risks and maximise the advantages of groups by gathering a massive amount of people without getting the attention of state security. As the prayers finished around the same time, it was also easier to carry out the marches in co-ordination with all other subgroups. Disciplined crowds exiting the mosques spread into the streets. They were shouting and raising their hands in an outburst of anger and energy following the distribution of leaflets advising on tactics, slogans and targets. The mosques enabled mass participation in the protests and facilitated the co-ordination of the process by providing a starting point for the protesters. The leaflets, on the other hand, led different groups to follow the same tactics and slogans (Beaumont et al. 2011). They thereby facilitated the harmony of the collective action.

Co-ordinating the protests from the mosques also helped to include different strata in the collective action. The biggest difference of 28 January protests from the first three days was the participation of a vast range of protesters coming from different segments of the society including the poor. As my interviewee Masud,[27] chief editor of the Ikhwan web, revealed:

> The ones who made the revolution successful were the poor people and the Islamists. The poor were the ones who demolished and burned the police stations in Egypt. They were against repression, poverty and corruption. They burned the police stations and in this way, they stopped the state's security structures from working. The Islamists mobilised the square, they resisted in the square and they fought with the state police until they won. That was how the Egyptian revolution became successful.

Masud actually refers to the Muslim Brotherhood in mentioning "Islamists" in his statement. Known as a fighter for the poor, the Muslim Brotherhood had created good networks in rural areas over the years (Munson 2001: 498). They thus became an important actor that mobilised and co-ordinated the poor segment of society in 2011.

On the other hand, the determination and power of the poor derived from their frustration and inability to carry on with their lives. Historical and sociological studies show that collective action is subject to the calculation of advantages and risks (Olson 2002: 368). As the poor segment had less to lose, they more effectively became involved in the collective action. The mobilisation and co-ordination of this segment mostly occurred via face-to-face communication (Tufekci and Wilson 2011). As mentioned in Chapter 3, the organisers went to different neighbourhoods to reach members of the public who did not have access to the Internet. To activate and mobilise them during the protests, they went to the streets and called them out with the promise of bringing the economic and political change they had long desired. Social movement theorists often argue that organising and mobilising people who have little economic power and few resources is tough (Greene 2005). In the Egyptian case, the protesters achieved this goal by including the expectations and demands of the working class in the slogans of the movement. Along with the slogan, "the people want to overthrow the regime", protesters chanted, "bread, freedom, social justice" on the streets (El-Ghobashy 2011). The Muslim Brotherhood was also an important actor that mobilised and co-ordinated the poor segment of society.

Internationalisation Tactics

The shutdown also became ineffective in the media via the collective efforts of journalists, NGOs, political parties and activists in Egypt. Despite the blackout, some of my interviewees, such as the blogger Wael Abbas, the human right activist Ahmed Samih, the journalist Wael Eskandar and the editor of Ikhwan web, Masud, stated that they broke the barriers of the state and spread the news outside of Egypt. Masud[28] described how they bypassed the communication cut:

> When the Internet was shut down, the young people who were good at mobilisation used some accounts, phone services and some other secret lines from Saudi Arabia. With these phones, we mobilised the young Ikhwans in Saudi Arabia and other parts of the Gulf. At that time, I was communicating with a person from the Brotherhood in the United States with that kind of mobile phone. He was the one who was managing the website. Ikhwan was not out of the picture during this shutdown, it was just stationed outside.

Masud's words exhibit that the blackout could not obstruct the activities of the oldest and largest Islamic organisation in Egypt, Muslim Brotherhood. The young cohort of the organisation was mobilising activists in different countries and informing international media about the events. On the other hand, the prominent blogger Wael Abbas, who constructed established networks through his blog on the basis of his past protest experiences, was also amongst the protesters who were little affected by the Internet cut. Wael[29] stated that:

> I was disconnected from the Internet for about three days, but most parts of the country were more disconnected than that. We had connections that the government was not aware of. Connections were providing us services to the stock exchange and private sector organisations' networks. We had used them without the government noticing. When the government learned about these, they completely shut down the Internet. We then used dial-up connections. I also used a satellite connection and regularly updated people.

Wael's statement exhibits the significance of having dial-up and satellite connections for protesters. When the state completely shut down the Internet by blocking the service of the Noor Internet service provider, the only option left to the protesters was satellite and dial-up connections.

The activists started securing their connections with the outside world through international media channels as well. International media connected with the activists via landlines too. Founded by human rights activists, the Hisham Mubarak Law Centre became a hub and meeting place for Egyptian protesters who reported the news to the international media (Rogin 2011). During our interview, the Egyptian activist Sherif Alaa,[30] for instance, noticed that:

> We were based at the human rights organisation called the Hisham Mubarak Law Centre. They were hosting the activists because this place was only 15 min away from Tahrir square and it was safe. So, we could go there and come back easily. The office had three or four rooms and we used to spend nights there. This centre became the focal point between Tahrir square and the international media, like the BBC, France 24, Al Jazeera, CNN and ABC. Most of the domestic and international news channels were calling us through landlines.

Sherif's comments are important to illuminate the vital roles of the organisations during the protests. The protesters broke the media censorship by using different communication methods, such as satellite

Internet, landlines and the stock exchange network's connections, but while using these methods they benefitted from the resources of established Egyptian organisations.

An alternative communication method was the dial-up system proposed by Jacob Appelbaum. The system was set up by Alaa Abdel Fattah, his wife Manal Hassan and sister Mona Seif. It aimed to keep communications alive during the shutdown imposed by Mubarak (Della Ratta and Valeriani 2012). Google also offered a new programme to Egyptians and converted their voicemail messages into text messages and posted them on Twitter. Sherif also added that fax machines were crucial during the protests and that activists sent messages with 140 characters via fax and that activists abroad distributed these messages on Twitter.[31]

The Self-protective Phase of Mubarak

On 28 January, Egypt experienced the most violent day of demonstrations in the history of Mubarak's 30-year rule. Police stations were attacked and the state experienced its biggest failure on the ground. Events then accelerated after Friday prayers. As security structures were demolished, the state tried to maintain order by announcing a nationwide curfew and deployed the tanks that surrounded the city (Kirkpatrick 2011). Nevertheless, while the Syrian army responded to the protests in that country by opening fire on the protesters, in Egypt, the army preferred not to interfere with the protests. During our interview in 2011, Masud,[32] the editor of Ikhwan web, defined the role of the army in this way:

> Two years ago, people were talking about what would happen if Gamal took power. Intellectuals were saying the army would not like it and they would overthrow him. The army could accept Mubarak as leader, but they would not accept his son, who was a civilian, educated in an American university and whose mother was a foreigner. That was against the army culture and against the military's rule. That's why the army hesitated to support Mubarak.

Decision-making processes in Egypt were highly centralised around the president over the terms of Sadat and Mubarak. However, as Italian sociologist Alberto Melucci (1996: 248) writes, "the state apparatus is never merely a docile, monolithic instrument in the hands of the dominant group". When the dominant interests of the army clashed with those of Mubarak, this centralised structure underwent dissolution amongst complex institutions (Salem 2013). The repressive apparatus of the state then cracked more easily in this centralised structure and could not re-establish the limits of the state.

In this section, I will analyse the unavailing efforts of Mubarak to re-establish stability. After three days of protests, Mubarak decided to take further steps in the mass media. He appeared on TV and announced that he had dismissed the government.[33] During this speech, he adopted a modern standard of Arabic that wasn't used in the Egyptian streets. Mubarak thus preferred a formal language that was mostly criticised by the experts for disconnecting him from the people ("Hosni Mubarak's speech" 2011). When his speech is analysed, one can see that Mubarak mostly dedicated it to a biographical narrative. Felix Berenskoetter (2012) defines biographical narratives as narratives by which communities define themselves in time and space. For instance, states often reflect a biography of their history, character and aspirations. In his speech, Mubarak also often presented a vision of his character and aspirations. In order to have a solid identity narrative, actors need to connect their past, present and future selves and create a feeling of biographical continuity (Berenskoetter 2012: 18). In his speech, Mubarak followed this rule and created biographical continuity by referring to his past and present as a leader. He described himself as a tolerant and liberal leader who had always ascribed importance to freedom of expression and the rule of law. He showed that he had always been on the side of his people by claiming that he had dissolved the government for the sake of the Egyptian people. The socio-cognitive approach, which reproduced justifications and denials of inequality in society, also featured in Mubarak's speech (Dijk 1993: 263). Mubarak accepted that the Egyptian economy needed to be fixed, but refused to take responsibility for its collapse. Rather, he blamed the existing government for the economic problems and created a new plan in which the government would be changed but he would still hold power.

A biographical narrative was also frequently used in Mubarak's speech to offer necessary information about where the nation was and where it was going. Such a vision can be positive or negative, underlining an undesirable world (Berenskoetter 2012: 18). Mubarak chose to use a negative vision and offered a dystopia. In this dystopian vision, Mubarak described what could happen if the country was left in the hands of the protesters and narrated an image that should be avoided. He called the protests terrorist activities that would bring chaos to the country. He therefore tried to guide Egyptian perceptions of the protests and maintain political stability.

Mubarak also tried to benefit from the glorious national biography of Egypt and reminded the public that the country had a distinct position in the region as the biggest country. He claimed that there would be no democracy in the Middle East if Egypt permitted chaos. He therefore called for the public to act responsibly and help to build a new future together. He finished his talk by describing himself as the father of the nation and promised to protect Egypt and Egyptians. Mubarak, thus, used different identity narratives, such as the nation's biography and his own, in different sections of his speech to try to shape understanding of the protests.

In order to maintain power over the discourse, the state also launched a hostile campaign against satellite stations that jeopardised stability and security (Freedom House 2012). It commenced this campaign by cancelling the network licence of Al Jazeera and shutting down its bureau in Cairo ("Egypt shuts down Al Jazeera bureau" 2011). Journalists on the ground also reported attacks and threats against them for destabilising the country. Mubarak then spread the word that the foreign media had sought to destabilise the country via the state media, particularly the state-run Al Nil TV. In line with Berenskoetter's theory (2012: 18), institutions in Egypt not only enabled some voices and their representations to become dominant, but also functioned to sustain the narrative of Mubarak.

Despite all these efforts by the state on the streets and in the mass media, the impacts of Mubarak's narrative on Egyptians were limited. In the subsequent days of 28 January, the protests continued and grew. For instance, on 1 February, the biggest protest since the protests began on the 25 January took place. More than a million people rallied in Cairo. Protestors took to the streets and called on Mubarak to step down in cities including Sinai, Alexandria, Sues, Mansoura, Damanhur, Arish, Tanta and El Mahalla El-Kubra ("Protesters flood Egypt streets" 2011). To restore stability, Mubarak gave his second speech on that day ("Hosni Mubarak's speech" 2011). It could be seen that in this second speech, Mubarak was more emotional and anxious about the situation that was unfolding in the country. This time, he altered his discourse and claimed that the protests were not organised by the Egyptian people but by groups who wanted to threaten stability and provoke the nation. He also adopted a new narrative about who the protesters were. He differentiated the protesters from the rest of the nation and referred them as "they" while he referred to the nation in terms of "us". As explained by linguist Teun A. Van Dijk, justification of the inequalities was conducted with two strategies: positive representation of the own group and a negative representation of the others. In Mubarak's second speech, these positive and negative representations were highly predominant. Mubarak related the uprising to an external threat that only pursued its own interests rather than the good of Egypt. On the other hand, he described Egyptians as anxious about tomorrow and their future.

Yet, what was striking was that unlike in his first speech, Mubarak described a future Egypt without him. He promised the public that he would not stand in the next elections and would do everything for his nation to transfer power to an authorised and legitimate government. In this way, he tried to gain the trust of the public again and influence their attitudes by representing himself as the defender of the rule of law and democracy. Mubarak also promised that he would bring justice to the country before the next elections and ensure that those who engaged in corruption would be punished. Yet, he did not forget to threaten those who were on the streets by saying that anyone who committed looting or started fires would be punished.

The final words of Mubarak adopted an emotional discourse and tried to touch the hearts of Egyptians. He said "I am proud of all the long years I have spent in the service of people of Egypt… This dear country is my country, just like it is the homeland of every Egyptian man and woman". Once more, he adopted a biographical narrative in which he described himself as a leader who emanated from his nation and who would die for his nation and people.

The apparent shift in the state's actions was reflected in the virtual arena as well. The high level of attendance of the protests showed that the media blackout did not meet the expectations of the state and, rather, induced a widespread reaction from the international community. The state found a solution by partially restoring the Internet and mobile service connection ("Internet service restored" 2011). The act facilitated the communication of regular Internet users. The new strategies adopted by the state would disrupt order in Tahrir Square, the focal point of the 2011 protests. Yet, as will be seen in the next section, they would not be sufficient to restore stability and order under Mubarak's rule.

Sustaining Peaceful Protests

The majority of my Egyptian interviewees referred to Mubarak's speech and mentioned that this emotional address to the nation was effective in touching the hearts of Egyptians. The Egyptian activist Radamis,[34] for instance, said that:

> When Mubarak did his speech on TV, people got very emotional. He (Mubarak) said, "I was born in this land and I will die in this land. I will not escape like Ben Ali". Lots of people in Tahrir Square thought this was true and they then left.

Radamis was asked what helped the protesters sustain the protests across 18 days and, while answering, he referred to the day when they struggled the most to keep protesters in the street. The reaction of the protesters reflected a decline in the movement's power. As social movement scholars often mention, the ideology of a social movement is crucial to its sustainability. In the context of this ideology, it is always possible to identify, more or less explicitly, a social actor against whom the movement must struggle (Melucci 1996: 351). In the Egyptian case, the different ideological groups unified around a non-ideological aim and only identified a social actor against whom they would struggle. The main slogan of the protests was "people want the downfall of the regime". Yet, for most people, the regime equated to Mubarak.[35] When Mubarak pointed out that he would not run in the next elections, for some protesters, the goal that the protesters were fighting for had been achieved. With the promise of Mubarak leaving politics, apart from the Egyptian revolutionaries who tried to keep the culture of dissent alive, others did not see any necessity to remain on the streets anymore. The undesirable situation that had activated the public was resolved. Moreover, the public no longer wanted either the clashes or the instability, as this was having a negative impact on the economy. As my interviewee, the Egyptian activist Sherif Alaa[36] revealed, many were called by their families and asked to return to their homes.

In the virtual realm, a fluttering of pro-Mubarak vemes (virtual memes) also spread fast. Vemes are a term that was coined by Linda Herrera and refer to "ideas whose conception, birth and proliferation occur inside virtual space" (Herrera 2014: 117). They differ from messages spread through traditional media with the incredible speed at which they move and the volume of people that they reach during their journey in virtual space. Mubarak "e-militias" were quick to spread these pro-regime vemes and infiltrated the opposition spaces on Facebook (Herrera 2014: 119). Sami (pseudonym),[37] the third admin of the Kullena Khaled Said Facebook page, described the reactions that they received on the page:

> We saw an immense ground campaign to discredit the revolution. I remember that there were massive attacks on our Facebook page as well. We were aware of the propaganda, which many people believed. Subsequently, I created a new protest event on Facebook for the next Friday. But, instead, half a million of the participants signed in for another event called "no protest". It was very disappointing for me because this was like a counter revolution. But it was very difficult to explain this to people. I was receiving accusations claiming that I was living abroad.

Sami's statement reminds me the fact that social media is a double-edged sword. As well as encouraging political participation by the public, it could also shake the rigidity of collective action by revealing divergences amongst group members. In this virtual space could be observed both Egyptian revolutionaries who tried to sustain the culture of dissent and counter-revolutionary forces that used Facebook to control and obstruct opposing voices (Herrera 2014: 117). The counter-revolutionary voices became particularly effective when revolutionaries operated anonymously in virtual space. Sami was indeed in Cairo while the protests were unfolding and actively participating in the protests on the ground. Yet, as he was operating anonymously in the online sphere, Egyptians who would follow his online activities believed that he was another soft leader who tried to mobilise the public from abroad. As Sami could not create the necessary bonds between him and the Internet users in Egypt, the counter-revolutionaries could easily manipulate user behaviour and thwarted the possibility of change. This side of virtual space would become particularly apparent after the revolution, when Internet militias appeared in virtual space and attacked prominent protesters' accounts.

In order to retain the popularity of the KKSFP, Sami[38] said that he responded to the activities of the counter-revolutionaries with the use of wit and sarcasm:

I found that confronting the propaganda did not work so well, so I adopted a humorous discourse. The square was full of humour. It was unbelievable. I was filming all of them, taking a lot of pictures and videos and then posting them on the Kullena Khaled Said page. If it was a humorous post, people would really like and share it. So, I continued with this method. I also posted comments from famous names that were supportive of the revolution. I also mostly referred to the economic difficulties that people faced during the Mubarak era. Members of the page also helped me design posters and suggested ideas.

The words of Sami show that the horizontal structure of the web enabled professional veme-makers like him to benefit from the creative ideas of page members. These example of Sami supports the theories of Herrera, who mentions that the counter-revolutionaries had difficulty competing with experienced veme-makers on the Internet (Herrera 2014: 117). Mubarak militias were mostly ineffective, since they used vemes that were largely inspired by the Obama 2008 presidential campaign and these did not fit the lifestyle of Egyptians. On the other hand, the revolutionaries had long-circulated pro-revolution vemes that displayed youth lifestyles in Egypt (Herrera 2014: 119). As a result, while the revolutionaries could create close connections with Egyptian Internet users, the state continued to lose control over its digital ideological apparatus.

What particularly aided the revolutionaries and maintained the stability of the collective action were also the government's blunders on the ground. On 2 February 2012, a former speaker of Egypt's upper house of parliament organised the so-called camel battle. During this, some of the regime supporters came to the Tahrir Square riding camels and horses and they attacked anti-government protesters with clubs, rocks, knives and firebombs ("Egypt speaker" 2011). The attack caused the death of several people and had an unexpected impact on the collective action, with the situation on the ground once more turning to the advantage of the protesters (Fahim and Kirkpatrick 2011a). The repression indeed amplified the Egyptian protests and helped the revolutionaries sustain the collective action in Egypt. The attacks showed the activists "how desperate and out-of-touch Mubarak's regime had become" (Tufekci 2017: 21).

Before the camel attacks began, information about possible attacks had already spread in the square. Sami[39] who was in the square on that day thought that these messages were mostly for provocation. He said:

I was on the street and nothing was going on, but I heard a lot of rumours that Tahrir was being attacked and it was very dangerous. I went home and wrote on our page, "Nothing is going on in the square; you can come from these and those streets". When I arrived at Tahrir, it was very crowded, I had to fight to get into the square. I remember that, during that fighting, I really thought if we stayed there, we would die. I was telling my friends, "I think we should go home and come back later". My friends told me, "we were the ones who asked people to come here, so we shouldn't leave them now". I remember that my last post on the page before going to the square was "everything is okay in Tahrir".

Sami's words actually reflected the possible damage caused by collecting information from social media. As Wael Eskandar[40] mentioned, during the Egyptian protests, journalists were following the news through Facebook and Twitter via the accounts of their trusted sources and seeking correct information from them. As Castells (2012: 58) claims, social media provided a technological platform for protesters to rise as trendsetters in the movement. However, the restricted Internet connection in Tahrir prevented the diffusion of instant information outside of the square. For instance, Wael indicated that the journalists who took photos in Tahrir had to come back to the media centre to upload their photos. While they travelled from the square to the centre, they missed the events that occurred in this period.

Revolutionaries also realised that one way to sustain the movement was to gain the support of international actors who would put pressure on the Egyptian state and weaken its power. For this aim, human rights activists, social movements and political organisations worked collaboratively to contact the international community. The Muslim Brotherhood, for instance, was one of the main organisations that provided information and visual data to international agencies. By November 2010, the Muslim Brotherhood had two websites with 10,495 pages (Howard and Hussain 2013: 61). During our interview, the editor of these pages, Masud,[41] said:

After 28 January, the Muslim Brotherhood's apparatus started to work. The young members of the Ikhwan were going outside and putting their photos on RASDDetect.com. We were monitoring the messages and photos of young Ikhwans on this web page. I was also supported by the Ikhwan, Brotherhood portal. They were following and monitoring all the news. We supported people who participated in the protests and offered them Internet access points. We contacted news agencies and informed them about the protests.

The efforts of Masud and other young Muslim Brotherhood members showed that the organisation was as significant in the virtual sphere as on the ground. They connected the protesters with the outside world, easing access to international media channels to help them get their messages heard. They therefore improved the monitoring, news gathering and distributing processes through their Internet web page, database and portal.

The availability of a large and vibrant civil society in Egypt, with 40,000 registered NGOs, also helped the sustainability of the collective action (Mikhail 2014). In their inspiring book, "Why Occupy a Square?", Gunning and Baron (2013) particularly emphasise the role of youth-focused organisations such as the Al Andalus Institute for Tolerance, Anti Violence Studies, Nahdet El Mahrousa (empowering young professionals) and Alashanek ya Balady (youth helping in socially deprived areas) for collective identity in Tahrir Square. These organisations were helpful for infusing democratic practices in the square, such as a blossoming of an urban, "non-sovereign" politics. In this urban politics, virtues such as tolerance between Muslims and Christians were the key elements of political life. Thanks to these spatial practices, groups from different ideologies and religions coexisted in Tahrir and collaboratively confronted the state. They created a humanitarian collective identity.

The NGOs were also critical for the sustainability of the movement. For instance, Horytna, the online radio station of the "Al Andalus Institute for Tolerance", assisted with the protests from the first day. Horytna conducted live reporting from Tahrir Square until the Egyptian government shut down the Internet. Mohammed,[42] editor and presenter of Horytna, mentioned that with respect to the shutdown of the communication structure, they had to stop live reporting. However, they had not left the square for 18 days and actively participated in the protests. He described what they did during these 18 days:

> We used the mobile phones and did live streaming by tools such as "Bambuser" (an interactive live video broadcasting service). We made many live programmes that were recorded in the square and uploaded them to our website, hurriyetna.com. We also conducted interviews with the people in the square and took lots of photos of every moment. We then uploaded them to our website and social media accounts, such as Facebook and Twitter. When we lost the Internet for a week, most people uploaded their videos later.

As Della Porta claims, social movements have in fact been described as weak players in the mass media sphere (Della Porta 2013: 89). However, ICTs helped protesters to overcome these difficulties in digitally supported protests. The vibrant civil society of Egypt and the opportunities that the new media provided enabled the protesters to frame the news and share it online and with the outside world. Unlike Syria, international media coverage of the protests in Egypt helped the protesters to sustain the dynamism of collective action (Lynch et al. 2014: 11). This was apparent with the changing official pronouncements that adopted a more supportive position towards the demonstrators, their civil rights and legitimate claims for democracy (Cottle 2011: 654).

The State Against a Utopia

After the camel battle of the 2nd of February 2011, the growing number of protesters in Tahrir Square pushed key members of the embattled ruling party, the NDP, to resign. The Vice President, Omar Suleiman, also tried to restore the state's legitimacy by opening a communication platform with opposition leaders (Sandels 2011). The return of the ruling party did not satisfy the protesters and millions in Tahrir continued to demand the resignation of President Mubarak. They transformed Tahrir Square into a revolutionary space, which challenged the disciplinary institutional order (Gunning and Baron 2013: 241). My interviewees described this space as a utopia in which ideological and sectarian tensions were left outside. Groups from different religions and ideologies prayed together, lived together and created a living space that belonged to all Egyptians. According to my female interviewees, such as Nisrin,[43] gender was also suspended in the square. She described the scene in Tahrir:

> I felt as if I was not a woman in the square. I was in Tahrir standing unveiled and no-one was looking at me. There were women from different backgrounds and ideologies. One minute, you could see a rich, affluent lady and then you could also meet a woman who was very conservative. On the other side, there were students as well. These gave the movement more and more credibility.

Nisrin's words support the description of Tahrir Square by scholars Gunning and Baron (2013), who describe it as "a political space that did not contain the traditional trappings of a sovereign space, instead displaying the multiplicity of self-governing qualities of urban politics" (Gunning and Baron 2013: 244). Thanks to this multiplicity, people from different ideologies, genders and religions could easily be integrated into this space and considered it their own.

The spirit of the protests was particularly boosted by the labour strikes. As was mentioned above, the workers had previously participated in the protests individually. However, with the re-opening of the factories, disparate worker groups came together and started a national strike (Knell 2011). While most of them demanded better salaries, many others also asked Mubarak to step down and abolish the emergency law (Beinin 2011). In cities such as Assiut, farmers blocked the main highway and railway to Cairo and in Port Said, slum dwellers set parts of governor's headquarters on fire ("Egypt's labor strikes" 2011). However, what brought power to the protests in particular was the intervention of the strategic groups of organised workers such as the Public Transport Authority Workers, postal workers and Suez Canal Company workers. These worker movements initiated the final phase of the revolution. They forced the army to break with Mubarak by stopping life in small, but also in big, cities (Gunning and Baron 2013: 181). The repertoire of workers played an important role in the success of this strike. Since 1998, more than two million workers from nearly every sector have participated in protests and mobilised large numbers for their cause (Beinin 2011). Thanks to their past protest experience, they knew well how to stage collective action and shift power between the protesters and the state. The protests had even been seen in the state-run *Al Ahram* newspaper. This government flagship newspaper suffered a significant setback and started to cover the attacks against protesters under the title, "offense to the whole nation" (Fahim and Kirkpatrick 2011b). The journalists of *Al Ahram* also challenged the government by asking for irreversible constitutional and legislative changes and by demanding better wages and more independence from it (Blomfield and Michaels 2011).

In order to regain its power over discourse, the state intensified its attacks on the protesters. The journalists had been hounded by state forces since the 28th of January, but the attacks escalated after the 2nd of February when Internet access was restored. International journalists, such as the *Washington Post's* Cairo Bureau chief Leila Fadel and *Canadian Globe and Mail* reporter Sonia Verma, were taken into custody. Attacks on the international correspondents of ABC, BBC, CBS and CNN were also reported (Journalists attacked and detained in Cairo, 2011). Acting as leaders, revolutionary youth attempted to create a dialogue between the protesters and the state. They sent a delegation to negotiate with the military and asked for a presidential council made up of civilians and officers and a government of national unity (Cole 2014: 161).

The last strategic move of Mubarak was his appearance on TV, in a speech to the nation on the 10th of February ("Egypt unrest" 2011). In this final speech, Mubarak changed his narrative. He directly addressed the youth and instead of describing the protesters as terrorists, he welcomed them as his sons and the youth of Egypt. He also represented the protesters as symbols of a new generation and a change for the better. Hence, the attempts of Mubarak to create social cognition, in which the protests would be seen as terrorist acts, were shattered by himself. According to Dijk (1993: 257), social cognition involves the influence of knowledge, beliefs, understandings, plans, attitudes, ideologies, norms and values. Contrary to his previous speeches in which he had threatened protesters or accused them of creating chaos, Mubarak embraced the protesters and blamed those who applied pressure on them, then promised that the bloods of the victims would not be forgotten. However, the changing narrative of Mubarak prevented it from having the intended effect and managing "social cognition" (Dijk 1993: 257). The large masses that continued to occupy the square and chanted "down with the regime", demonstrated the continuing negative perception and beliefs towards the Mubarak government.

Finally, Mubarak tried to create a dystopia in which an external threat against Egyptians from outside Egypt was claimed. As Berenskoetter (2012: 12) argues, "dystopias can be used as warnings". By offering an image of "what could happen if", they narrate a being-in-the-world that should be avoided. By spreading scare stories about an external factor, Mubarak used the dystopia to warn the public. He also once more adopted an emotional discourse and reminded people that he had served his country for 60 years. This identity narrative described himself as a responsible and loyal man and protector of the transition process. This was the first time that Mubarak described the protesters as victims too.

The changing discourse of Mubarak did not work on the Egyptians though, they had been expecting the departure of the president on that night, and so the last speech of Mubarak induced mostly anger and disappointment. Along with Tahrir Square, the protesters invaded key political points, such as the front of the presidential palace and the state television building (Fahim and El-Naggar 2011). They also tried to show Tahrir was not under the sovereignty of the state. In this space, purified from the police forces, the protesters succeeded in creating their own media sphere by distributing a fake version of Al Ahram, in which pro-revolutionary sentences and mockeries gained attention (Fahim and El-Naggar 2011). New initiatives, such as the Tahrir Square Facebook page,[44] were also launched. The protesters distributed information from the ground to the international media via this (Fahim and El-Naggar 2011). Moreover, the protesters succeeded in activating each segment of the society against the state on the ground. This, in return, left few options open to the army, which would oust Mubarak and replaced him with Omar Suleiman on 11 February 2011.

Conclusion

Tracing the mobilisation, co-ordination and sustainment of the Egyptian protests demonstrated that although the Internet was accessible to only 35 per cent of the Egyptian population in 2011,[45] it was crucial for baffling the police and enabling the mobilisation crowds without police intervention during the mobilisation process. My interviews with the protesters also showed that action was based on mass participation, but there was also a centralised decision-making structure with a clear division of labour, similar to the professional movement organisations (Jenkins 1983: 539). The organising group, the Revolutionary Youth Council, secured the centralisation and coherence of protest activities by working together and planning the entire mobilisation process. The past protest experiences and networks of this group helped them use the online and offline communication methods effectively. Besides effectively benefitting from the ICTs, they also used offline communication methods such as word-of-mouth and face-to-face communication to mobilise the protesters. In addition, they created new strategies on the ground, such as using small streets as gathering points and distributing information about protest sites via leaflets and brochures. These findings support the theories of communication scholars who focused on the Egyptian protests, such as Aouragh and Alexander (2011), who argue that there was a dialectical relationship between online and offline political action during the 25 January protests. In addition to this argument, the interviews conducted for this research revealed that the protesters were supported by traditional political organisations such as members of the Muslim Brotherhood and experienced activists who individually participated in the protests. This outside support enhanced the technological facilities of the group and enlarged their networks.

When the co-ordination of the protests is analysed, the activists used ICTs as an effective resource in this process as well. Starting from different protest sites, they co-ordinated with the help of mobile phones, social media, Google maps and SMS. The protests were well planned by experienced activists, hybrid leaders and soft leaders in advance and the capital of the country did not lose touch with the peripheries. Even the shutdown of the Internet did not interrupt the mobilisation and co-ordination of the protests. As the protesters had prepared in advance for a possible Internet cut, they managed to commence the protests from different protest sites in the absence of ICTs. They used the mosques as starting points and followed Google map plans that they prepared in advance. Human rights organisations such as the Hisham Mubarak Centre, as well as political groups such as the Muslim Brotherhood and experienced activists facilitated the work of organising groups and secured a connection with the outside world. Researchers who analysed the relation between civil society and the ICTs argue, for instance, "The ICTs give civil society new tools to support their claims" (Laer and Aelst 2010: 230). My interviews with Egyptian protesters show that ICTs not only support the claims of civil society, but that civil society also shaped the way in which ICTs were used during the protests. ICTs were effectively used in Egypt by the Muslim Brotherhood and other political organisations. These civil society groups were not part of the leadership in Tahrir Square, yet they were important for the co-ordination and internationalisation of the movement. Thanks to their human and technological resources, as well as past experiences, they effectively used the ICTs to connect the protesters to the outside world, as well as to co-ordinate the movement.

I have also found that sustaining the protests was one of the most challenging aspects for the activists. Different ideological groups had unified with the aim of overthrowing Mubarak on 28 January. When Mubarak mentioned that he would not run for elections on that night, this unification almost cracked and anti-revolutionary messages spread in both the offline and online spheres. Current research on the protests and ICTs mostly analyses the role of ICTs in building the movement and co-ordinating it. Yet, sustaining the movement was as important as mobilising and co-ordinating it. Tufekci (2017) was one of the few scholars who refer to the sustainability of digitally supported movement in her book, *Twitter and Tear Gas*. She criticises digitally networked movements for withering away without having an impact on policy. To some extent, the Egyptian case proved this argument of Tufekci, as it showed that ICTs might have a double-edged effect: they also served the interests of the counter-revolutionaries and had damaging effects on the protests, such as downgrading the words of the social media administrators. As the administrators operated anonymously, this also created further cynicism in the public, decreasing their trust and empathy. Yet, the Egyptian protesters managed to reverse this situation and regain the competitive advantage by responding to attacks with wit and sarcasm on the Internet. The hybrid leaders took an active role in this process and challenged the anti-revolutionary forces by spreading sarcastic messages online that criticised these forces.

Using ICTs was one way to challenge the anti-revolutionary forces, yet it was not enough to sustain the protests on the ground. What particularly brought support to the Egyptian protests was, again, the repression coming from the state. After the camel battle, the number of protesters in Tahrir Square increased. Tolerance was another important factor that sustained protesters from different ideologies in the square. The past protest experience of the activists, the presence of human rights organisations and exposure to social media amplified Egyptians' level of tolerance and empathy (Howard and Hussain 2013). Hence, Egyptian protesters could succeed in sustaining their movement and not withering away due to their past protest experiences, but also through the vibrant civil society that surrounded them, which supported the protests and provided facilities for the protesters. The civil society, which consisted of human rights organisations, worker movements and established political organisations, played an important role in shaping the way that ICTs were used and in determining the direction of the Egyptian protests.

All in all, confirming Tufekci's argument, this chapter showed that in the absence of necessary resources, a digitally supported movement can easily wither away without having an influence on policy (Tufekci 2017). My findings also proved the continuing relevance of resource mobilisation theory in today's digitally supported movements. Aside from the ICTs, resources such as the past experience of the protesters, their networks, outside support from organisations and state repression levels all played an important role in the sustainability of the Egyptian protests.

NOTES

1. Interview with Sherif Azer, Egyptian activist, London, 2014.
2. http://carnegieendowment.org/2013/11/14/violence-against-copts-in-egypt-pub-53606.
3. Interview with Abdelrahman Ayyash, Egyptian activist, Istanbul, 2013.
4. Interview with Mona (pseudonym), Egyptian activist, Cairo, 2014.
5. Interview with Mona (pseudonym), Egyptian activist, Cairo, 2014.
6. Interview with John Albert, Egyptian activist, Alexandria, 2014 (refer to appendix).
7. Interview with Masud (pseudonym), Egyptian journalist, Ikhwan web editor, Cairo, 2011.
8. Interview with John Albert, Egyptian activist, Alexandria, 2014.
9. Interview with Ahmed (pseudonym), Egyptian activist and researcher, Cairo, 2011.
10. Interview with Youssef (pseudonym), Egyptian blogger, Cairo, 2014.
11. Interview with Rana (pseudonym), Egyptian activist and journalist, Cairo, 2014.
12. Skype interview with Mohammed, Egyptian activist, 2015.
13. Interview with Abderrahman Ayyash, Egyptian activist, Istanbul, 2013.
14. Interview with Wael Eskandar, Egyptian journalist, Cairo, 2014.
15. Interview with Wael Eskandar, Egyptian journalist, Cairo, 2014 (refer to appendix).
16. Interview with Mona (pseudonym), Egyptian activist, Cairo, 2014.
17. Interview with Ahmed Samih, Egyptian activist, head of the Andalus Institute for Tolerance and Anti-Violence Studies, Cairo, 2014.
18. Interview with Sherif Azer, Egyptian activist, London, 2014.
19. Interview with Sherif Alaa, Egyptian activist, London, 2013.
20. Interview with Sherif Azer, Egyptian activist, London, 2014.
21. Interview with Wael Abbas, Egyptian blogger, Cairo, 2014.
22. Interview with Wael Eskandar, Egyptian journalist, Cairo, 2014.
23. Interview with Mona (pseudonym), Egyptian activist, Cairo, 2014
24. Interview with Esraa Abdel Fattah, Egyptian activist and blogger, Cairo, 2014.
25. http://www.mei.edu/profile/basem-fathy.
26. Skype interview with Mamdouh Salah Eldeen, Egyptian activist, 2014.
27. Interview with Masud (pseudonym), Egyptian journalist, Ikhwan web editor, Cairo, 2011.
28. Interview with Masud (pseudonym), Egyptian journalist, Ikhwan web editor, Cairo, 2011.
29. Interview with Wael Abbas, Egyptian blogger, Cairo, 2014.
30. Interview with Sherif Alaa, Egyptian activist, London, 2013.

31. Interview with Sherif Alaa, Egyptian activist, London, 2013.
32. Interview with Masud (pseudonym), Egyptian journalist, Ikhwan web editor, Cairo, 2011.
33. https://www.youtube.com/watch?v=9DtOr6BBOHg.
34. Interview with Radamis Hany Zaky, Egyptian activist, Cairo, 2011.
35. Interview with Radamis Hany Zaky, Egyptian activist, Cairo, 2011.
36. Interview with Sherif Alaa, Egyptian activist, Cairo, 2014.
37. Interview with Sami (pseudonym), Egyptian activist, Cairo, 2014.
38. Interview with Sami (pseudonym), Egyptian activist, Cairo, 2014.
39. Interview with Sami (pseudonym), Egyptian activist, Cairo, 2014.
40. Interview with Wael Eskandar, Egyptian journalist, Cairo, 2014.
41. Interview with Masud (pseudonym), Egyptian journalist, Ikhwan web editor, Cairo, 2011.
42. Interview with Mohammed, editor of Horytna radio station, Cairo, 2014.
43. Interview with Nisrin (pseudonym), Egyptian activist, London, 2013.
44. https://bit.ly/2TL8H0Q.
45. https://bit.ly/2FLJQ9E.

References

Adi, A., Gerodimos, R., & Lilleker, D. (2018). 'Yes we vote': Civic mobilisation and impulsive engagement on Instagram. *Javnost—The Public, 25*(3), 1–18. https://doi.org/10.1080/13183222.2018.1464706.

Aouragh, M., & Alexander, A. (2011). The Arab Spring: The Egyptian experience: Sense and nonsense of the Internet revolution. *International Journal of Communication, 5,* 1344–1358.

Beaumont, P., Shenker, J., & Black, I. (2011, January 28). Egypt protesters defy curfew as tanks roll into Cairo. *The Guardian.* Retrieved from http://www.theguardian.com/world/2011/jan/28/egypt-protesters-tanks-cairo.

Beinin, J. (2011, February 17). Egypt's workers rise up. *The Nation.* Retrieved from http://www.thenation.com/article/egypts-workers-rise/.

Berenskoetter, F. (2012). Parameter of a national biography. *European Journal of International Relations, 20*(1), 262–288. https://doi.org/10.1177/1354066112445290.

Blomfield, A., & Michaels, A. (2011, February 7). Egypt's crisis: Hosni Mubarak lost control of state media. *The Telegraph.* Retrieved from http://www.telegraph.co.uk/news/worldnews/africaandindianocean/egypt/8309469/Egypt-crisis-Hosni-Mubarak-loses-control-of-state-media.html.

Castells, M. (2012). *Networks of outrage and hope: Social movements in the Internet age.* Cambridge, UK: Polity Press.

Cole, J. (2014). *The new Arabs: How the millennial generation is changing the Middle East.* New York, NY: Simon & Schuster.

Cottle, S. (2011, July). Media and the Arab uprisings of 2011, Research notes. *Journalism, 12*(5), 647–659. https://doi.org/10.1177/1464884911410017.

Della Porta, D. (2013). *Can democracy be saved? Participation, deliberation and social movements.* Cambridge, UK: Polity Press.

Della Porta, D., & Diani, M. (2006). *Social movements: An introduction.* Oxford, UK: Blackwell.

Della Ratta, D., & Valeriani, A. (2012). Remixing the spring! Connective leadership and read-write practices in the 2011 Arab uprisings. *Cyber Orient, 6*(1). Retrieved from http://www.cyberorient.net/article.do?articleId=7763.

Dijk, V. T. (1993). Principles of critical discourse analysis. *Discourse & Society, 4*(2), 249–283. https://doi.org/10.1177/0957926593004002006.

Egyptians report poor communication services on Day of Anger. (2011, January 25). *Egypt Independent.* Retrieved from http://www.egyptindependent.com/news/egyptians-report-poor-communication-services-day-anger.

Egypt's labor strikes break out across the country; protestors defiant. (2011, September 2). *Huffington Post.* Retrieved from http://www.huffingtonpost.com/2011/02/09/egypt-labor-strikes-break_n_820909.html.

Egypt shuts down Al Jazeera bureau. (2011, January 30). *Al Jazeera.* Retrieved from http://www.aljazeera.com/news/middleeast/2011/01/201113085252994161.html.

Egypt speaker 'plotted battle of the camel'. (2011, July 14). *Al Jazeera.* Retrieved from http://www.aljazeera.com/news/middleeast/2011/07/2011714172156277964.html.

Egypt unrest: Full text of Hosni Mubarak's speech. (2011, February 10). BBC. Retrieved from http://www.bbc.co.uk/news/world-middle-east-12427091.

El-Ghobashy, M. (2011). *The praxis of the Egyptian revolution.* Middle East Report 258: People Power, 41. Retrieved from http://www.merip.org/mer/mer258/praxis-egyptian-revolution.

Eltantawy, N., & Wiest, J. (2011). Social media in the Egyptian revolution: Reconsidering resource mobilization theory. *International Journal of Communication, 5,* 1207–1224.

Eskandar, W. (2015, January 28). The significance of Shaimaa's death. *Daily News Egypt.* Retrieved from https://bit.ly/2FRrvZk.

Ezzat, A. (2014, September 29). 'You are being watched!' Egypt's mass Internet surveillance. *Mada Masr.* Retrieved from http://www.madamasr.com/opinion/politics/you-are-being-watched-egypts-mass-internet-surveillance.

Fahim, K., & Kirkpatrick, D. D. (2011a, February 2). Mubarak's allies and foes clash in Egypt. *The New York Times.* Retrieved from https://nyti.ms/2L0YOeD.

Fahim, K., & Kirkpatrick, D. D. (2011b, February 9). Labor actions in Egypt boost protests. *The New York Times.* Retrieved from http://www.nytimes.com/2011/02/10/world/middleeast/10egypt.html.

Fahim, K., & El-Naggar, M. (2011, February 8). Emotions of a reluctant hero galvanize protestors. *The New York Times*. Retrieved from http://www.nytimes.com/2011/02/09/world/middleeast/09ghonim.html?_r=0.

Freedom House. (2012). *Egypt: Freedom of the press 2012 report*. Retrieved from https://freedomhouse.org/report/freedom-press/2012/egypt#.VdXtMrJViko.

Gerbaudo, P. (2012). *Tweets and the streets: Social media and contemporary activism*. New York, NY: Pluto Press.

Ghonim, W. (2012). *Revolution 2.0*. Boston, MA: Houghton Mifflin Harcourt.

Greene, J. (2005). Whatever it takes: Poor people's organising, OCAP, and social struggle. *Studies in Political Economy, 75*, 5–28.

Gunning, J., & Baron, I. Z. (2013). *Why occupy a square: People, protests and movements in Egyptian revolution*. New York, NY: Oxford University Press.

Herrera, L. (2014). *Revolution in the age of social media*. London, UK: Verso.

Hofheinz, A. (2011). The Arab Spring nextopia? Beyond revolution 2.0. *International Journal of Communication, 5*, 1417–1434.

Hosni Mubarak's speech: Full text. (2011, February 2). *The Guardian*. Retrieved from http://www.theguardian.com/world/2011/feb/02/president-hosni-mubarak-egypt-speech.

Howard, P., & Hussain, M. (2013). *Democracy's fourth wave? Digital media and the Arab Spring*. New York, NY: Oxford University Press.

Internet service restored in Egypt. (2011, February 2). *Al Jazeera*. Retrieved from http://www.aljazeera.com/news/middleeast/2011/02/201122113055781707.html.

Jenkins, J. C. (1983). Resource mobilisation theory and the study of social movements. *Annual Review of Sociology, 9*, 527–553.

Journalists attacked and detained in Cairo. (2011, February 3). *Independent*. Retrieved from https://bit.ly/2Htkpu0.

Khamis, S., & Vaughn, K. (2012). Cyberactivism in the Egyptian revolution: How civic engagement and citizen journalism tilted the balance. *Arab Media & Society, 14*. Retrieved from http://www.arabmediasociety.com/articles/downloads/20120313094800_Khamis_Cyberactivism_updated.pdf.

Khamis, S., Gold, P. B., & Vaughn, K. (2012). Beyond Egypt's "Facebook revolution" and Syria's "YouTube uprising": Comparing political contexts, actors and communication strategies. *Arab Media & Society, 15*. Retrieved from http://arabmediasociety.com/articles/downloads/20120407120519_Khamis_Gold_Vaughn.pdf.

Kirkpatrick, D. (2011, January 28). Mubarak orders crackdown, with revolt sweeping Egypt. *The New York Times*. Retrieved from http://www.nytimes.com/2011/01/29/world/middleeast/29unrest.html?pagewanted=all&_r=0.

Knell, Y. (2011, February 14). *Egypt strikes test military regime*. BBC. Retrieved from http://www.bbc.co.uk/news/world-middle-east-12460657.

Kuebler, J., & Allagui, I. (2011). The Arab Spring and the role of ICTs. *International Journal of Communication, 5*, 1435–1442.

Laer, J. V., & Aelst, P. V. (2010). Cyber-protest and civil society: The Internet and action repertoires in social movements. In Y. Jewkes & M. Yar (Eds.), *Handbook of Internet crime* (pp. 230–254). Cullompton, UK: Willan Publishing.

Levinson, C., & Coker, M. (2011, February 11). The secret rally that sparked an uprising. *The Wall Street Journal.* Retrieved from http://www.wsj.com/articles/SB10001424052748704132204576135882356532702.

Lynch, M., Freelon, D., & Sean, A. (2014, January). Syria's socially mediated civil wars. *United States Institute of Peace Report No. 91.* Retrieved from http://www.usip.org/sites/default/files/PW91-Syrias%20Socially%20Mediated%20Civil%20War.pdf.

McCarthy, J. D., & Zald, M. N. (1977). Resource mobilization and social movements: A partial theory. *American Journal of Sociology, 82*(6), 1212–1241.

Melucci, A. (1996). *Challenging code: Collective action in the information age.* Cambridge, UK: Cambridge University Press.

Mikhail. A. (2014, October 6). *The obliteration of civil society in Egypt.* Open Democracy. Retrieved from https://www.opendemocracy.net/arab-awakening/amira-mikhail/obliteration-of-civil-society-in-egypt.

Miskimmon, A., O'Loughlin, B., & Roselle, L. (2013). *Strategic narratives.* New York, NY: Routledge.

Munson, Z. (2001). Islamic mobilisation. *The Sociological Quarterly, 42*(4), 487–510.

Olson, M. (2002). *The logic of collective action: Public goods and the theory of groups.* Cambridge, MA: Harvard University Press.

Osman, A., & Samei, M. A. (2012, Spring/Summer). The media and the making of the 2011 Egyptian revolution. *Global Media Journal, 2*(1), 1–19. Retrieved from http://www.db-thueringen.de/servlets/DerivateServlet/Derivate-25453/GMJ3_Samei_final.pdf.

Papacharissi, Z., & Oliveira, M. (2012). Affective news and networked publics: The rhythms of news storytelling in Egypt. *Journal of Communication, 62*(2), 266–282. https://doi.org/10.1111/j.1460-2466.2012.01630.x.

Protesters flood Egypt streets. (2011, February 2). *Al Jazeera.* Retrieved from http://www.aljazeera.com/news/middleeast/2011/02/20112113115442982.html.

Protests in Egypt-Wednesday. (2011, January 26). *The Guardian.* Retrieved from https://www.theguardian.com/world/blog/2011/jan/26/egypt-protests.

Rogin, J. (2011, February 2). White House failing to consent Mubarak to start transition now. *Foreign Policy.* Retrieved from http://foreignpolicy.com/2011/02/02/white-house-failing-to-convince-mubarak-to-start-transition-now/.

Salem, S. (2013, September 6). The Egyptian military and the 2011 revolution. *Jadaliyya*. Retrieved from http://www.jadaliyya.com/pages/index/14023/the-egyptian-military-and-the-2011-revolution.

Sandels. A. (2011, April 2). *Egypt: Mubarak-controlled state TV expands coverage of protests* [Web log post]. Retrieved from http://latimesblogs.latimes.com/.m/babylonbeyond/2011/02/egypt-state-tv-cover-protests-organizer-.html.

Tilly, C. (2006). *Regimes and repertoires*. Chicago, IL: University of Chicago Press.

Tufekci, Z., & Wilson, C. (2011). Social media and the decision to participate in political protest: Observations from Tahrir Square. *Journal of Communication, 62*(2), 363–379. https://doi.org/10.1111/j.1460-2466.2012.01629.x.

Tufekci, Z. (2017). *Twitter and tear gas: The power and fragility of networked protest*. New Haven and London: Yale University Press.

Wilson, C., & Dunn, A. (2011). The Arab Spring: Digital media in the Egyptian revolution: Descriptive analysis from the Tahrir data sets. *International Journal of Communication, 5*, 1248–1272.

CHAPTER 5

Breaking the Silence: The Efforts of Syrian Activists to Organise and Mobilise Digitally Supported Protests

As with the Egyptian revolution, the seeds of the Syrian protests were sowed on the Internet. Inspired by the protests in Tunisia and Egypt, three Syrian friends outside of the country decided to take action for Syria and launched a Facebook page called, "The Syrian Revolution 2011" (Aslan 2015). Two years later, I got the opportunity to interview one of the founders, Mahmoud (pseudonym) in a small café in London. During our interview, Mahmoud emphasised the influence of Tunisia and Egypt on them. While the 2011 protests were unfolding in Tunisia and Egypt, like many Syrians, Mahmoud and his friends were following the news with great interest.[1] After all, the few protests that Syrians *had* witnessed since the 1980s were anti-Syrian protests that took place in the Kurdish regions in 2004 (Kurdwatch 2009) and regime-sanctioned protests on foreign issues. The success of their Tunisian and Egyptian counterparts induced Mahmoud and his friends to think that the time was right for a change. They chose Facebook as their main protest tool, as this digital platform provided them with a new dimension of power that they had previously lacked: digital connectivity.

The advantage of Facebook was that the administrators acquired the chance to reach distant friends, colleagues, acquaintances with whom they had had weak ties and normally had little chance to stay connected to. More importantly, as their messages were shared by a person with whom they had a weak tie, they could also reach out to the friends and acquaintances of this person. Like their activist counterparts in Tunisia

© The Author(s) 2020

B. Aslan Ozgul, *Leading Protests in the Digital Age,*
Palgrave Studies in Young People and Politics,
https://doi.org/10.1007/978-3-030-25450-6_5

and Egypt, they would thereby be able to reach friends, friends of friends and the networks of friends of friends of friends, etc. Hence, the messages would rapidly spread from one cluster to another on the back of digital connectivity.

However, unlike most of the cases analysed by communications scholars, the power of digital connectivity could not help the three friends ignite a protest in their first attempt and pushed them to adopt alternative strategies. In this chapter, I analyse the organisation and mobilisation attempts of the Syrian activists and disclose communication methods utilised by the activists to organise the protests. Their answers also enlighten how and why particular desires and methods resonated in their various local contexts (Aslan 2015). I also explore whether the three types of leaders, noticed in the Egyptian protests, were present in the Syrian protests.

So far few studies explain why the communication tactics of the Syrian protesters show difference from those of their counterparts in other Arab countries. By investigating the network society of Syria in 2011, Enrico de Angelis (2011) supports the argument that Syria's Internet space emerged after the protests broke out in Syria, unlike Egypt's well-developed Internet community. It was developed by a number of newly active and disconnected actors, mostly operating from overseas. It is precisely due to its infancy that Syria's Internet activism appeared incompetent of either leading the protests or presenting a platform where dissidents could discuss a unified political stance. This chapter develops Angelis' argument through in-depth interviews with the Syrian Revolution 2011 Facebook page (SRFP) administrator, as well as Syrian protesters on the ground.

The main finding of this chapter is that variations in regime types and particularly in protesters' repertoire of contention affect the way in which they used the technology. As it was explained previously in the history chapter, due to the desire of stability of Syrians and the power of the regime in enforcing and sustaining its fictions Syrians preferred to distance themselves from political life. When the Syrian uprising started in 2011, most Syrians were still inexperienced in staging a demonstration, creating social ties and identities. This chapter first shows how this lack of a repertoire of contention affected the use of ICTs in Syria. Analysing the activities of the "soft leaders"—the administrators of social media networks—show that to activate a public who did not engage in political action was only possible by creating trust relations amongst the participants and by establishing notions of causality and belonging. Due to the heavy hand of the state in the society, the Syrian administrators did not have experience in leading a social media page before 2011. Related to this inexperience in using ICTs for political purposes, the actors failed to create trust relations with the followers over the communication on the Internet. The chosen date for the protests, the spokesperson with a Muslim Brotherhood background and the religious language used by the administrators further complicated attempts to overcome identity boundaries amongst the participants. Interviews with Syrian protesters indicate that what motivated the Syrians to join the protests was the successful mobilisation of Tunisian and Egyptian protesters, the grievance that arose in Daraa, the social networks formed on the ground over years and the intense repression towards the protesters.

Syria's Soft Leaders and Their Online and Offline Activities

As explained in the introduction and Chapter 3, today's digital movements are often described as crowd-enabled connective actions in which ICTs replaced the role of social movement organisations and leaders, leading to the emergence of a new style of protest. In this new type of protest, "... most tasks were taken care of by horizontal organisations that evolved during the protests or by unaffiliated individuals who had simply shown up" (Tufekci 2017: 51). The starting point of the Syrian protests matched well with the dynamics of the crowd-enabled connective actions. For instance, the idea of a Syrian revolution was first announced on a Facebook page and the three administrators of this page attempted to mobilise the public via digital media tools. Yet, a crowd-enabled and digitally connected protest logic would not work in Syria. In the state-controlled public sphere of Syria, in which the presence of the state could be felt in the country's political, social and public life (Saleh 2003: 65), starting a protest on Facebook was not a simple and easily undertaken task. As we were discussing the organising process of the 2011 Syrian protests, Mahmoud,[2] the administrator of the Syrian revolution page, reminded me of the state-controlled political sphere in the country and the difficulty of launching a public page on Facebook called, "Syrian Revolution". He said:

> We launched the page but it was very scary, even for us... The Syrian
> Embassy in Britain and all other countries knew me because I had pro-
> tested in front of Syrian embassies from 2005/2006. Talking about the
> protests in front of the embassies and reading statements had been
> accepted until now but talking about a revolution against the regime was
> not acceptable. This would have an impact on me, on my family inside the
> country and on all my relatives.

Mahmoud felt the necessity of explaining to me Syrian political culture
to ensure that I comprehended how Syria was differed from Egypt in
terms of state-control in the political sphere and how afraid Syrians were
to engage in political action. The dissidents believed that talking about
the regime had serious consequences so they were cautious to express
their opinions online, not only for their own safety but also for the safety
of their relatives. My interviewees also said they were afraid of speaking about politics on social media before the revolution. Bilal,[3] a Syrian
dissident who was living and studying in Damascus when the uprising started, stated, "there were no politics in our posts, but there were
human issues, there was fear, tension, sadness and frustration".

Unlike the Egyptian protesters, who had been blogging, tweet-
ing and posting their political ideas since 2004, Syrians feared expressing and sharing their political opinions on social networks. My Syrian
interviewees stated that they had even been cautious about saying the
name of the Syrian President, "Assad", in their houses. They warned
each other, "the walls have ears".[4] After the start of the Syrian uprising,
most Syrian protesters broke this fear and exposed news about the
uprising, although some still feared to post on social media. After all,
the Internet sphere of Syria was subject to strict control by the authorities, as with other media channels. For years, cyber-dissidents and bloggers who tried to express their political opinions had to write using an
alias to avoid being arrested or tortured (Ghrer 2013: 7). Whoever did
not concede the greatness of the leader and referred to taboo subjects
like "politics, ideology, religion, society, economics and discussions of
Syria or the Middle East" was subject to arrest (Cooke 2007: 9). In an
attempt to protect its dominance in the online sphere, the state launched
a wave of blogger arrests as soon as the Tunisian and Egyptian protests
erupted. On 20 August 2011, for instance, blogger Ahmad Hadifa,
known by the blog name of Ahmad Abu Al-Kheir, was arrested in

Baniyas by military security officials (Reporters without Borders 2011). A teenage blogger, Tel Malouhi,[5] who had been arrested two years previously, was also sentenced to five years' imprisonment on the basis of espionage charges (Zackheim 2011).

On the ground, although Mahmoud and his friends joined the protests outside of Syria, protest attempts were rarely seen inside the country. It was also not possible to find a well-established civil society that consisted of movements, civil organisations and online activists in Syria (Angelis 2011).[6] Intolerance was the main characteristic of the state, which was decisive in keeping the political domain fettered (Ghrer 2013: 117). It repressed the movements of intellectual forces, which appeared during the 1990s and 2000s (Kilo 2011: 161). These intellectuals, who played a controversial role in the production of Syrian culture, were "silenced if they cannot be heard outside the environment that suppresses and co-opts them" (Cooke 2007: 164).

In this state-controlled political sphere, being abroad motivated the three founders of the SRFP. They also knew that if they operated online, they did not need to reveal their identities offline. For instance, although it has been three years since the beginning of the Syrian uprising, Mahmoud still prefers to stay anonymous and uses his alias in interviews. For that matter, he chooses not to appear on TV channels. When asked why he prefers to use an alias outside the Syria, he stated:

> For security reasons... It is not only about Syrian intelligence, but about all intelligence agencies from around the world. If the intelligence agencies of other countries know your identity, they will not allow you to stay in their country. Many Syrians get deported because of that, so we have to protect our identities. (Aslan 2015)

The precautions that Mahmoud has taken were also adopted by the other admins of the page. In this unsafe environment, the Internet arose as the only platform permitted to operate anonymously against the state.

Apart from being the first open platform for communication between activists and the organisation of protests, the SRFP played an informative role. For instance, the information on the time and location of protests that had started to happen in Syria in February 2011 was released via Facebook. The interviewees described this call as a very courageous and shocking act. Bilal,[7] for instance, told me that:

> Before the revolution started, if you remember, there had been a trial in February. There was a call for a protest from the Syrian Revolution page on Facebook. When I saw the first online post, it was a shock. The first thing I did was showed it to my family members. I said, "Look what is going on" and then, there was this silent moment… (Aslan 2015)

Bilal's remarks were signs of the fear and inexperience of Syrians when it came to protests. Seeing a call to action that directly defied the authorities surprised Syrians at first and caught their attention. Yet they could not instantly return to this call, as they feared the potentially atrocious reaction of the state. Many other interviewees for this research also stated emotions of shock and surprise, similar to those uttered by Bilal. Some interviewees, such as Nadia,[8] reported that the initial call on 5 March was seen as a hoax, a trick by the Syrian security services. Thus, albeit staying anonymous enabled the administrators to work in safety, it thwarted their ability to earn the trust of people. Syrians who were doubtful about the reliability of the Facebook page did not respond to the call to action received from the page immediately. Mahmoud, the administrator of the Syrian revolution page, pointed out that the TV and press media had a fundamental part in the reputation and increased membership of the page (Aslan 2015). He noted that:

> In the first two weeks, we were writing things like "please help us to reach 5,000". We had to be 5,000 people. We kept writing this post for 6-7 hours and some brave-hearted people invited their friends and we got support from other Arabs. Arabs from other countries were the ones who liked the page in the beginning. When we got 5,000 likes, a lot of media outlets noticed our page and started to write about it. AFP, a French news agency, wrote a report about it and this report was published in 100-200 newspapers around the world because AFP reports are widely diffused. The BBC made a report about the page for two or three minutes. The following week Al-Jazeera, the Arab Qatari TV, also did a report about us. Within three weeks, we became like a star and everybody heard about the page.

The words of Mahmoud illustrate the significance of traditional media in increasing the prominence of Facebook pages and convincing people of the credibility of social media. While the rise of the Internet has reduced the role of traditional media in framing protests, TV channels still play an important role in digitally supported protests. As Almqvist

(2013: 60) claims, without the mainstream media, the Internet would only be a logistical network for core activists and a public sphere for members of the Syrian diaspora. The TV channels that look into mediating stories from inside Syria became the mediator between the public and the admins of the social networks. After all, despite recent developments in the capabilities of ICTs, few Syrians could access the Internet. To reach non-Facebook users, organisers relied on TV channels. Moreover, most of the old activists who participated in past protests in Syria did not have social media accounts. Traditional media was also useful in connecting the small numbers of activists on the ground to the admins of social media channels.[9] The closed interconnections of traditional and digital media facilitated the organisation process and made the admins heard by the Syrian public. This interconnection between the Internet and traditional media shaped the evolution of the Syrian Revolution Facebook page—an interdependence still constitutive of media systems in the West, too (Chadwick 2013: 188).

However, as explained in the beginning of this chapter, despite the admins' intensive efforts, the first protests in Syria organised via social media did not spark a revolution. No one even turned up for the digitally supported action three administrators called for. Essentially, there were many reasons behind this failure. According to Ribal Al-Hassad, cousin of Bashar and director of the London-based Syrian organisation for democracy and freedom, the protest's date and the photos used on the Facebook page created reservations in people's minds. The administrators picked a date that mirrored the date of the Muslim Brotherhood uprising in Hama in 1982. Al-Hassad also adds that the illustration used on Facebook, of a clenched fist and red colour, evocative of blood, resembled a calling for a civil war, which Syrian people were not looking for (Lesch 2012: 92). Besides, the religious language that the administrators adopted in their posts was criticised by certain segments, who saw this as an attempt to whip up sectarian hysteria (Aslan 2015; Nabki 2011: 5). One of the main reasons for this mistrust of the public was that the SRFP was only founded on 18 January, only a few months before the protests erupted. Thereby, it was a new platform for the public and did not have enough time to build trust amongst the people. A spokesperson, Fida'ad-Din Tariif as-Sayyid, who agreed to speak on TV on behalf of all of the other admins, became the only connection between the admins and the public. Mahmoud mentioned that they used this religious language without any sectarian aim; they just wanted to communicate in the everyday language of the Syrian people. They later convened a consultation board to consider the content of their communication, but this was challenging process because the consultants did not always respond quickly.[10] Hence, the wrong narrative choice, insufficient time to generate trust and the security caution of admins all had an effect on the low level of attendance at the protests.

The hesitance of the people to attend these protests fortify theories of diffusion and mass communication studies, which claim that people seldom act on mass media information if it is not conveyed through personal ties (Katz and Lazarsfeld 1955). Contrary to the crowd-enabled connective action logic, where unaffiliated masses could join the protests after getting news from ICTs, the Syrians did not want to engage in protests with people with whom they had weak ties, who they did not know well. As the admins did not establish personal ties with the society, the Internet by itself was not sufficient to convince people. Manual Castells (2013) attempts to explain this problem, noting that coping with fear is a fundamental prerequisite for people wishing to get involved in collective action. In order to overcome fear, the emotional mechanism of togetherness is first required (Castells 2012: 10). In the limited period between the preparation and ignition date of the protests, the administrators did not have sufficient time to construct these ties with the society. Another challenge was that most of the Syrian people in Damascus lacked protest experience (Aslan 2015). It is important to remember that "there is a logic of collective action that entails relational structures, the presence of decision-making mechanisms, the setting of goals, the circulation of information, the calculation of outcomes, the accumulation of experience and learning from the past" (Melucci 1989: 17).

Factors such as accumulation of experience and learning from the past were missing elements in the Syrian protests. Without past protest experience, Syrians preferred to wait for others to take the first initiative. Mazen,[11] for instance, a Christian activist and journalist from Homs, told me:

> My friends and I attempted to go to the protests, but then a couple of us said that no-one was going and that it was very dangerous, so we did not go.

Clearly, being active offline and being active online are not same things and have different risks. Offline activism brought a lot of adversities with it and in the lack of a well-organised group; dissidents were scared to take the risk of protesting against the oppressive Syrian state (Aslan 2015). As Zeynep Tufekci (2014) argues, instead of looking at the outputs of social media-enabled protests, studies today should also explore the role of digital media in capacity-building. The Syrian uprising shows that the Facebook page alone was not sufficient for improving the capacity of the public. As explained in Chapter 3, to build trust relationships between page members, the administrators needed to adopt different offline strategies, such as organising events in advance at which the page members could enter into face-to-face interactions. A final factor that prevented the spark of protest was the absence of experienced activists in the protest area. Mahmoud[12] states that what the admins needed was actually the support of experienced protesters. He continued:

> At these times, you need pioneers to go and protest in the middle of Damascus. Some of my friends later told me that they went to the place of the protests, the one we called for. They went there to see… If somebody came, they would join. So, everyone waited for the person to come. We needed someone to ring the bell.

The arguments of Mahmoud are illustrative of the limited capacities of the admins in the organisation of protests. Due to the fear of the security services and absence of effective opposition, the organisers could not mobilise people by merely relying on hybrid media tools. They also had to draw on the dynamics of collective action.

The unsuccessful first protest attempt organised via the Internet forced the administrators of the Facebook page to incorporate additional tactics in the repertoire of contention, defined by Tilly (2006) as the subset of methods employed by people for making claims against powerful others. They benefitted from traditional methods, such as face-to-face communication with experienced activists. Mahmoud[13] claimed that:

> We once more called for protests on 15 March. We were prepared for it. We said that we were not going to wait for someone but that we would find this person. Thus, we began to communicate with activists and told them "you should go!" I had never been in jail, not even for an hour. I could not imagine myself in jail. Ex-prisoners did not know such fear because they had spent 15 years in prison. And they were not in English prisons, but in the most horrific prisons on earth.

The experience of Mahmoud poses the question of how actors construct their connective actions when challenging the legitimacy of power. Olson's (2002) collective action theory proposes that the simple premise of common interest is not sufficient to explain the mobilisation of protests. While taking action, protesters consider the rational estimate of the costs and benefits of an action. This theory explains the free riders amongst the Syrian people, who were afraid of the possible outcomes of the protests. Although the common interest could create togetherness in the online space, it could not be seen in the offline arena. In the absence of a well-organised opposition and experienced people who could lead the crowd, the public became reluctant to challenge the state. The experienced activists who had previous protest experience and who were ready to cope with the possible results of protests became the initiators of the spark in the offline sphere. They also gave courage to others. Thus, the 15 March protest, was organised via the co-operation of the Facebook administrators and old activists, became the first protest planned in Syria against the state which called for democratic reforms and the release of prisoners (Sinjab 2011). Until then, Syrians had only witnessed spontaneous protests, such as at a market in the Hariqa district in Damascus, which had been triggered by an altercation between the police and a market trader ("The revolution reaches Damascus" 2011). 15 March protest also received coverage by different media channels, such as the BBC and Al Jazeera. The videos of the protest quickly made their rounds on YouTube as well. However, when the attendance level was analysed, according to witnesses, approximately 50 people attended ("Protesters stage rare demo" 2011).

If the Facebook page played any role in the emergence of the Syrian uprising, it was particularly in its success in being the first public platform that encouraged the public to take action. The attempt by the admins was immediately taken as an example. On 16 March, 100 experienced activists and relatives of political prisoners gathered in front of the Interior Ministry to request the release of political prisoners. The seeds of the uprising were sown.

Power of Established Social Networks and Collective Identity

Although the admins of the SRFP were the first people who tried to mobilise the people, they were not the only ones who aimed to initiate change inside the country. Thanks to the images coming from Tunisia and Egypt, the fear was subsiding in the country and young people's casual discussions took on an unfamiliar, political tone (International Crisis Group 2011: 8). Inspired by the Egyptian and Tunisian protests, the youth started to talk with their close friends and relatives about the possibility of a revolution. Mizyan[14] was one of them; this software engineer has been waiting for change in Syria for years. He participated in the anti-Iraq War protests in Syria and had already been arrested twice by the state for his political activities. He told me that:

> In 2011, we followed the events in Tunisia and Egypt. We were excited about this, frankly! A friend of mine was in Tunisia and he came to Syria in February 2011. He talked about what they did there and what we could do for Syria. I told him, "We need to do something here!"

Like Mizyan, Syrians who were eager to initiate the protests in their hometowns started to talk with each other about freedom. Iyas,[15] for instance, was a tourist guide in Saraqib. During the uprising, he became well known as the painter of Saraqib's walls with graffiti about the desires of the Syrian people. He claimed:

> We were talking too much about the Arab Spring, about what happened in Egypt and what happened in Tunisia. We were talking about freedom but, because of the regime, we were very afraid, so we weren't talking loudly.

The statements of these dissidents from different cities of Syria show how Arab Spring protesters inspired Syrians. While the dictators toppled one by one in the region, the coffee shops became the social space of Syrians, where the public discussed social issues. All of these events illustrated that Syrians were breaking down the fear and creating social ties. The uprising was going to start sooner or later, even in the absence of the Daraa events, but would it suddenly bring millions to the streets? Unlike previous, small-scale protests in Damascus and Aleppo, the protests that sparked in the provincial town of Daraa succeeded in bringing millions to the streets. What was the difference between the Daraa protests and the 15 March protests organised by the admins of the SRFP? Why did one spark the revolution while the other only brought 50 activists to the streets?

The Daraa protests were not planned like the 15 March protests. In early March 2011, inspired by the Egyptian and Tunisian protesters, a number of youth drew anti-government graffiti on the walls of a school. They were arrested by the Moukhabarat and their fingernails were pulled out. When the parents asked for the return of their children, they were told "you will not be getting your children back. Go home and tell your useless husbands to make you some more" (Starr 2012: 3–4). For most of the parents, who were from prominent clans with respectable networks, this was unacceptable. The words attributed to the local representative humiliated the honour of the clans. Within three days, the people of Daraa took to the street. The crowds who invaded the streets after the Friday prayers in the Ummari Mosque on 18 March were the first initiators of the massive uprising. As they protested, a few were shot dead (Castells 2012: 100; Leenders 2012: 428).

The scholar Reinoud Leenders (2012) explored the dynamics and underlying conditions of the Daraa uprising by analysing digital sources and contacting activists on the ground. While explaining the factors that ignite the protests in Daraa, Leenders draws attention to the opening political opportunities. According to Leenders (2012), the well-established social network of Daraa citizens, including clans, labour, migration, cross-border movements and crime, was one of the reasons behind the successful mobilisation of the Daraa uprising. This network was particularly established by the cross-border trade between Daraa and Jordan (Leenders 2012: 426). In addition, there was a criminal network in the city that generates more networks, skills, tactics, resources and social spaces essential for the mobilisation process (Leenders 2012: 426). These findings of Leenders show that social networks, which formed on the ground over the years, remain vital for the development of protests in authoritarian countries such as Syria.

Essentially, thanks to ICTs, activists can challenge de facto spokespersons online and create online networks over which they share similar interests in the online sphere. To spark a protest, they do not specifically need offline networks formed on the ground over years. However, as Syrians were afraid to express their thoughts, concerns and expectations in the online sphere, these offline networks remained crucial for the formation of a collective identity in the Syrian public sphere that would ignite protests. The prominent social movement scholar Alberto Melucci (1996: 63–64) argues that the free riding of protesters can be avoided by the formation of a collective identity—the formation of the "we" through which people recognise themselves, confer meaning and provide continuity. What was seen in the Daraa protests was in fact this collective identity. The appearance of the protests in Daraa was not a coincidence. It can be explained by the collective identity, which had been present in the city for years.

In addition to the closed social networks in the city, two other evident factors that distinguished Daraa protests from former protests in Damascus organised by the SRFP. First, the Daraa protests sparked spontaneously against the atrocious repression of the state, mediated by the overlapping social networks built over the years. The people witnessed the detention of 15 children and the death of peaceful protesters, respectively. Such serious violence and threat triggered the outrage, which gave way to the widespread protests. On the other hand, when we analyse the 15 March protests, organised by the SRFP, these were not acting in response to a political opportunity triggered by an immediate act of repression or threat from the state. As the researcher Enrico de Angelis asserts: "the Facebook page of the Syrian Revolution was not dedicated to a specific event, like the 'We are all Khaled Said Webpage' of Egyptian revolution and it was established by people who are not based in Syria".[16] It was more difficult for the admins to activate the emotions of people and mobilise them in Damascus, where security was greater than in other cities.

Another distinguishing factor between the Daraa protest and those in Damascus was the past protest experiences of the Daraa protesters. As an experienced activist, Osama, for instance, drew attention to the significance of past civil movements in mobilising citizens. Osama was also a participant in the 16 March protests in Damascus. When the Daraa protests began, he organised the protests in his home town of Darayya with Ghiath Matar, who was called as "martyr" by Osama. Matar was well-known for handing soldiers flowers and became the symbol of the peaceful protests. When Osama[17] was asked how they accomplished breaking down the fear in citizens and activating them in Darayya, he answered:

> The situation in Darayya was suitable for these things because, in 2003, there was a civil movement there. There were protests against corruption and the invasion of Iraq. So, the atmosphere was ready, and the people were prepared for the demonstrations there. In 2003, there were almost 50 young men who went to the protests and then the protests got bigger and bigger and it became 1,000 people. I had some friends who were arrested in 2003 in those protests and then released after two and a half years. They had big roles in mobilising people in 2011. (Aslan 2015)

Osama's remarks acknowledge that past collective actions acted a vital part in the mobilisation process during the 2011 uprising. The physical presence of members of past movements helped to form and sustain a collective sense of "we-ness" more rapidly (Earl and Kimport 2011: 125). This collective sense was present in cities such as Darayya, Daraa, Deir ez-Zor, where past protest rallies had taken place. The citizens knew how to collaborate with each other and they had already formed their repertoire of contention. Thus, the past protests experience of Daraa citizens and their well-developed social networks helped to overcome the obstacles to collective action. This connectedness also created a particular trust amongst the people (Aslan 2015). Trust was closely related to the motivations of the participants. As it augmented their motivation, it accelerated the mobilisation process too (Granovetter 1973: 1374). For instance, when I asked Abbas,[18] an activist from Daraa, what motivated people to participate in the Daraa protests, he emphasised the close bonds of Daraa's citizens, claiming:

> Daraa was the last city that French occupation had reached and it was the first city that French occupation left. The people of Daraa have a specific culture, they have strong ties. Our humanity, our feelings, our honour are important.

Abbas and Osama's remarks show that the factors that helped bring people to collective action, such as the repertoire of contention, social networks and cultural frames (Tarrow 2011: 33), had already been prepared in Daraa. These factors created confidence in the participants.

The catalysing factor, which mobilised all these groups, was also in particular the violent death of peaceful protesters during the protests in Daraa. Death has always been a significant factor that triggers violent emotions in people. It brings people together with little in common, but the grief of protesters creates solidarity amongst them (Tarrow 2011: 47). Abbas was one of these protesters. The 37-year-old doctor had had enough of the repressive regime and had been dreaming about the revolution since the Tunisian and Egyptian protests had erupted. He told me that when he heard about the protests in Daraa centre on 18 March 2011 he decided to arrange big demonstrations for the next day. He gathered his colleagues in Daraa hospital. He then encouraged them to participate in the protests and said they could not forget the blood of Wissam Ayyash and Mahmoud al-Javabra, who had been killed on that day in Daraa. During our interview in Istanbul, Abbas[19] described how they planned the 19 March protests:

> We called each other by phone and did not speak openly because the security service was monitoring the phones. Some friends were in the village. We called them and said, "Tomorrow, you and your friends should come to the square". All of my friends were ready the next day at 11 am. We went to the stationary shop. We opened the store quietly and bought fabrics and wrote slogans on them. We went to the mosque, did our prayers and yallah. We chanted, "Allahu Akbar, Allahu Akbar", "freedom, freedom", and "the blood of our martyrs won't be forgotten!" At first, we were 100 people. After the first half hour, about 700 people came. Later on, we were 1,000 and then more and more... The number grew.

Abbas's statement reveals that some of the activists who left the mosques also used the religious language, which created identity boundaries amongst the Syrian Internet users. However, the funeral of six Syrians who were shot dead on 18 March became a protest in its own right and activated Damascus, Homs and Baniyas as well as Daraa on 18 March ("In Syria crackdown after protests" 2011). It particularly gave way to massive anti-regime opposition demonstrations against the Ba'athist government on 19 March 2011. According to *The New York Times*, 20,000 people attended these demonstrations and the crowd asked for the removal of Syria's long-standing emergency law (Sterling 2012). The slogans of the protesters that referred to the martyrs, such as "the blood of our martyrs won't be forgotten", were influential and very effective in activating the citizens of Daraa ("Syrian police attack marchers at Funerals" 2011). After all, blood has always been a serious issue for the people of Daraa, who belong to tribes or big families. As Della Porta (2014) states, historical traditions are important in determining how protesters respond to repression. In this largely tribal society, family, loyalty and honour were important traditional values (Macleod 2011). In order to defend these, the society mobilised around the deaths. When the attempted repression by the government increased to intolerable levels and became a threat to citizens, the well-developed social networks in Daraa helped protesters forge solidarity with the parents of those children and made them feel that they were not alone. The ceremony then became an opportunity for the organisers of the protests to create a sustainable protest movement.

Conclusion

In the winter of 2011, Syrian citizens were excited and hopeful to see the Arab Spring protests spread to their region. The administrators of the SRFP aimed to tap into this excitement to mobilise a movement, becoming the first actors attempting to mitigate fear amongst the Syrian people. Along with the Internet, TV channels played an important role during the preparation process and were crucial for heightening prominence of the Facebook page and the administrators' credibility. This finding is corroborated by recent research, arguing that the initial nuclei of new collective action emerge in the form of online discussion (Gerbaudo 2012: 160).

Nevertheless, my findings also indicate that the lack of past protest performances prohibited the administrators from acquiring necessary online experience and they did not know how to mitigate fear amongst the public. Although the administrators started an online discussion in a constructed emotional space, due to an ill-conceived choice of narrative, a restricted time period and security fears, they failed to reconstruct a sense of togetherness and trust in the society. Consequently, the hybrid media activities of the administrators were not sufficient to convince large numbers of people to join the collective action. This finding goes against recent generalisations on Arab uprisings which argue that social media reconstructs a sense of togetherness (Castells 2012) and facilitates gatherings in the offline space (Tufekci and Wilson 2012). It is apparent that people still rarely act on mass media information in authoritarian states with a coercive security apparatus if this is not transmitted through personal ties.

The spontaneous protests in Daraa reveal that the components of a repertoire of contention such as the presence of social networks, the past protest experience of activists and shared grievances and aspirations in the city were still the main elements in forming and sustaining a collective identity, which was necessary for the emergence of sustainable collective action. Disproportionate state repression also created grievance that was a condition for the formation of solidarity amongst the public.

The next chapter will analyse the mobilisation, co-ordination and sustainability of the protests across Syria. It will also shed light on the activities of the state trying to manage the uprisings in a hybrid media context.

Notes

1. The majority of the Syrian interviewees also mentioned the emotions of enthusiasm and hope while describing their feelings about the Egyptian protests.
2. Interview with Mahmoud (pseudonym), the administrator of the Syrian revolution page, London, 2013.
3. Skype interview with Bilal Zaiter, Syrian activist, 2013.
4. Interview with Emad, Syrian activist, Istanbul, 2013.
5. http://talmallohi.blogspot.co.uk/.
6. See the history section for further information.
7. Skype interview with Bilal Zaiter, Syrian activist, 2013.
8. Skype interview with Nadia (pseudonym), Syrian activist, 2014.
9. Interview with Mahmoud (pseudonym), the administrator of the Syrian revolution page, London, 2013.
10. Interview with Mahmoud (pseudonym), the administrator of the Syrian revolution page, London, 2013.
11. Interview with Mazen, Syrian activist, Istanbul, 2014.
12. Interview with Mahmoud (pseudonym), the administrator of the Syrian revolution page, London, 2013.
13. Interview with Mahmoud (pseudonym), the administrator of the Syrian revolution page, London, 2013.
14. Interview with Mizyan Altawil, Syrian activist, Istanbul, 2013.
15. Skype interview with Iyas Kaadouni, Syrian activist, 2013.
16. Interview with Enrico de Angelis, researcher, Cairo, 2014.
17. Interview with Osama, Syrian activist, Istanbul, 2013.
18. Interview with Abbas (pseudonym), Syrian activist, Istanbul, 2014.
19. Interview with Abbas (pseudonym), Syrian activist, Istanbul, 2014.

References

Almqvist, A. (2013). The Syrian uprising and the transnational public sphere. In C. Wieland, A. Almqvist, & H. Naddis (Eds.), *The Syrian uprising: Dynamics of an insurgency* (pp. 47–78). Fife, Scotland: St. Andrew Papers on Contemporary Syria.

Angelis, E. (2011). The state of disarray of a networked revolution. *Sociologica, 3*, 1–24. https://doi.org/10.2383/36423.

Aslan, B. (2015). The mobilisation process of Syria's activists: The symbiotic relationship between the use of information and communication technologies and the political culture. *International Journal of Communication, 9*, 2507–2525.

Castells, M. (2012). *Networks of outrage and hope: Social movements in the Internet age.* Cambridge, UK: Polity Press.

Chadwick, A. (2013). *The hybrid media system: Politics and power.* New York, NY: Oxford University Press.

Cooke, M. (2007). *Dissident Syria: Making oppositional arts official.* Durham, NC: Duke University Press.

Della Porta, D. (2014). On violence and repression: A relational approach (The government and opposition/Leonard Schapiro memorial lecture, 2013). *Government and Opposition, 49*(2), 159–187. https://doi.org/10.1017/gov.2013.47.

Earl, J., & Kimport, K. (2011). *Digitally enabled social change.* Cambridge, MA: MIT Press.

Gerbaudo, P. (2012). *Tweets and the streets: Social media and contemporary activism.* New York, NY: Pluto Press.

Ghrer, H. (2013, April). Social media and the Syrian revolution. *Westminster Papers in Communication and Culture, 9*(2), 115–122.

Granovetter, M. (1973). The strength of weak ties. *American Journal of Sociology, 78*(6), 1360–1380.

In Syria crackdown after protests. (2011, March 18). *The New York Times.* Retrieved from https://nyti.ms/2JWKg0Z.

International Crisis Group. (2011). *Popular protest in North Africa and the Middle East (VI): The Syrian people's slow motion revolution.* (Middle East & North Africa Report No. 108). Retrieved from https://bit.ly/2FNSZP0.

Katz, E., & Lazarsfeld, P. F. (1955). *Personal influence: The part played by people in the flow of mass communications.* New York, NY: The Free Press.

Kilo, M. (2011). Syria… The road to where? *Contemporary Arab Affairs, 4*(4), 431–444. https://doi.org/10.1080/17550912.2011.630884.

Kurdwatch. (2009, December). The Al-Qamishli uprising: The beginning of a new era for Syrian Kurds? *European Centre for Kurdish Studies, 4.* Retrieved from http://www.kurdwatch.org/pdf/kurdwatch_qamischli_en.pdf.

Leenders, R. (2012, December). Collective action and mobilization in Dar'a: An anatomy of the onset of Syria's popular uprising. *Mobilization, 17*(4), 419–434.

Lesch, D. W. (2012). *Syria: The fall of the house of Assad.* Cornwall, UK: TJ International.

Macleod, H. (2011, April 19). Inside Daraa. *Al Jazeera.* Retrieved from https://bit.ly/1IYX2ps.

Melucci, A. (1989). *Nomads of the present: Social movements and individual needs in contemporary society* (J. Keane & P. Mier, Eds.). London, UK: Hutchinson Radius.

Melucci, A. (1996). *Challenging code: Collective action in the information age.* Cambridge, UK: Cambridge University Press.

Nabki, Q. (2011, May 2). *Talking about a revolution: An interview with Camille Otrakji* [Web log post]. Retrieved from http://qifanabki.com/2011/05/02/camille-otrakji-syria-protests/.

Olson, M. (2002). *The logic of collective action: Public goods and the theory of groups.* Cambridge, MA: Harvard University Press.

Protesters stage rare demo in Syria. (2011, March 15). *Al Jazeera.* Retrieved from https://bit.ly/2CMNfmu.

Reporters without Borders. (2011, February 25). *Ahmad Hadifa released and will not be charged.* Retrieved from https://bit.ly/2CNlnit.

Saleh, Y. (2003, March). The political culture of modern Syria: Its formation, structure & interactions. *Conflict Studies Research Centre, 57–68.* Retrieved from https://bit.ly/2UlUiNc.

Sinjab, L. (2011, March 4). *Syria: Why is there no Egypt-style revolution?* BBC. Retrieved from https://bbc.in/2HSOQvf.

Starr, S. (2012). *Revolt in Syria: Eye-witness to the uprising.* London, UK: Hurst & Co.

Sterling, J. (2012, March 1). *Daraa: The spark that lit the Syrian flame.* CNN. Retrieved from https://cnn.it/2Uig3gZ.

Syrian police attack marchers at funerals. (2011, March 19). *The New York Times.* Retrieved from https://nyti.ms/2uCnv83.

Tarrow, S. (2011). *Power in movement: Social movements and contentious politics.* New York, NY: Cambridge University Press.

The revolution reaches Damascus. (2011, March 18). *Foreign Policy.* Retrieved from https://bit.ly/2I9X4OS.

Tilly, C. (2006). *Regimes and repertoires.* Chicago, IL: University of Chicago Press.

Tufekci, Z. (2014, January 9). *Capabilities of movements and affordances of digital media: Paradoxes of empowerment* [Web log post]. Retrieved from https://bit.ly/1oqYgO2.

Tufekci, Z. (2017). *Twitter and tear gas: The power and fragility of networked protest.* New Haven, CT: Yale University Press.

Tufekci, Z., & Wilson, C. (2012). Social media and the decision to participate in political protest: Observations from Tahrir Square. *Journal of Communication, 62*(2), 363–379. https://doi.org/10.1111/j.1460-2466.2012.01629.x.

Zackheim, M. (2011, April 15). Syria's teenaged prisoners of conscience. *Al Jazeera.* Retrieved from https://bit.ly/2TW1sow.

CHAPTER 6

How the Peaceful Protests Turned into Armed Struggles in Syria

Even before the protests unfolded in Daraa's streets, journalists started to ask if "the dictatorship dominoes in the Arab and Middle Eastern world had begun to topple?" (CNN Wire staff 2011). The Syrian protests, however, would be unique amongst the Arab uprisings, as they were characterised by a wave of killings and arrests, torture and humiliation at the hands of the government. What started as a peaceful protest movement turned to violence, armed confrontation and eventually civil war (Sinjab 2011). Despite the growing number of research on Egypt, few have focused on the mobilisation and co-ordination process of the Syrian activists. Researchers such as Angelis (2011) and Della Ratta and Valeriani (2012) describe the Internet activism in Syria as uncoordinated, inexperienced and unable to provide accurate information about the protests to international media. As the peaceful protests in 2011 unfolded though, activists' use of ICTs showed progress. For instance, Della Ratta and Valeriani (2012) analysed the Syrian Internet almost one year and half after they began. They found that the online sphere had an important impact on the protests by opening up citizen forums and pushing forward the idea of an active citizenship over time. The Syrian protesters had also learned how to self-document the events they participated in and turned into first-person narrators of their own history (Della Ratta 2018). Yet, there were many more questions that crossed my mind: How was the fear amongst Syrians mitigated? Which methods did the activists used to mobilise the public in different Syrian cities? What were the roles

© The Author(s) 2020
B. Aslan Ozgul, *Leading Protests in the Digital Age,*
Palgrave Studies in Young People and Politics,
https://doi.org/10.1007/978-3-030-25450-6_6

of social and traditional media in this? In this chapter, by simultaneously tracing the course of events and the experiences of activists in the peaceful protests during 2011, I aim to provide an in-depth analysis of the actors and their progress in using ICTs and other communication strategies. I also address the relationship between online and offline political action to shed light on the organisational structure of the peaceful Syrian protests.

The data enable us to explore the leadership and mobilisation of local activists who mobilised the public, primarily through face-to-face communication and snowballing mobilisation tactics. ICTs, on the other hand, were mostly used for interpersonal communication. The reason why the online sphere in Syria improved in the uprising's later stage and affected the protests on the ground was the development of the repertoire of contention of Syrians. The eruption of protests in the authoritarian state mitigates fear in the society and led to the formation of identities and social ties. As explained in Chapter 4, this transition had already been completed in Egypt *before* the uprising started (Tufekci 2014a; Colla 2012). However, in Syria, despite the efforts of intellectuals in the 2000s, the repertoire of contention was still weak. As Tilly explains, repertoires vary from non-existent to weak, to strong, to rigid. The familiarity of previous performance and the likelihood that it will appear again indicates the different relationships of each position. In weak repertoires, past familiarity with protests increases the likelihood of subsequent performance and the protesters show the effects of learning, but still do not have strong preferences (Tilly 2006: 40). In the case of Syria, it was seen that experienced activists led the protests but did not have strong preferences at the start. As they became familiar with different protest techniques and created social ties with different groups, they also benefitted from hybrid media and built transnational public networks. Second, during the media blackouts, satellite phones and satellite Internet connections became the only communication tools of the protesters, connecting them with the outside world from their authoritarian countries. However, not everyone could access this privilege. The chapter indicates how certain dissidents who faced the endless challenge to generate media coverage fell short. Finally, this chapter shows how authoritarian states, such as Syria, try to maintain their systems of surveillance, censorship and limiting of discourse with the help of hybrid media. The state's media activities helped to convince the silent segment of the society that the sustainability of the state was essential for the sake of the country.

I start by analysing the reactions of the state to the appearance of the protests in Daraa and then continue by tracing the activities of the local activists whom struggled to adjust their activities according to the online and offline infrastructure imposed by the state.

First Reactions and Media Activities of the State

While the Daraa demonstrations appeared in March 2011 in Syria, the Syrian state was quick to manage the hybrid media environment, creating uncertainty about the details of the unrest and the exact number of protestors killed. Such a quick move from the state's side should not be surprising, as news media had always been under the control of the Assad regime in Syria (Al Barazi 2013). Under the state of emergency, the Press Law gave the state the right to control newspapers, books, radio and television broadcasting, as well as advertising and the visual arts (Pies and Madanat 2011: 11).

With the start of the uprising, the state's controls on the media tightened even more. Foreign reporters had been prevented from entering or working in Syria. The office of Al Jazeera was shut down, and the government threatened Al Jazeera staff and a web journalist, Dorothy Parvaz, was detained on her arrival at Damascus airport (Naggar 2011). All of these state precautions showed that, unlike in Tunisia and Egypt, the state enacted an effective media blackout from the start of the protests.

The government detained and assaulted journalists inside the country too ("Attacks on the press" 2011). The news media became the propaganda outlet of the state. Journalists from national media channels worked under strict control by the state. Rezan,[1] for instance, had been a journalist at one of the biggest Syrian national political newspapers. When the uprising started, she was still working at the newspaper at Damascus. She described the tough situation that her and other journalists in the Syrian media faced:

> I was very afraid of saying what I really wanted at work, because all of the employees had some relation to the security services. Each time they asked me what my opinion was about the uprising, I was afraid and stayed silent, because I knew what had happened to the journalists who told the truth. One journalist was killed by the Syrian army in his house as he worked with the revolutionaries. So, I was afraid. Once, an officer from the

> Moukhabarat asked me to come to his house. I told my husband that we had to leave the country immediately, as I was sometimes talking loudly without noticing and I was afraid that they had heard me.

Rezan would flee Syria that year and start living and working as a journalist in Turkey. Her fear demonstrates the effective monitoring system used by the state to control the activities of journalists. The fear, not only of getting arrested, but also about the unknown future, led to self-censorship and prevented independent journalism. Rezan mentioned that, over time, professional journalists like her who opposed the government's repression fled the country or were arrested, tortured and killed. Those who stayed in the country and continued to work were forced to broadcast government lies (Reporters without Borders 2012). For instance, in the early days of the uprising, state TV denied the storming of the mosque in Daraa. Instead, it said "an armed gang" had attacked an ambulance in the city. Security forces killed four attackers and wounded others; they also chased others, who fled (Mroue 2011). The discourse of armed gangs and attackers was adapted and repeated by the state authorities as well as media in the subsequent days. In the absence of independent media and foreign reporters, it was easier for the state to dominate the media narrative with its messages.

In the online realm, Facebook and YouTube have been blocked in Syria since 2007. Moreover, a number of political websites, such as news sites that issued articles critical of the Syrian government were also blocked. However, the state re-permitted Facebook and YouTube in February 2011. This could actually have been an attempt to more closely monitor activity on these sites (Heacock 2011). Alternatively, seeing the groundswell of protests across Arab countries, the state may have decided to keep its press freedom promises and to permit these websites. The aim was to appease the society and obstruct possible unrest in the country (Williams 2011). Yet, despite all the precautions by the state, the uprising erupted and gradually developed in Daraa.

At the beginning of the uprising, it was not possible to see any specific online tactics of the state. This might be due to the fact that it already tightly controlled ICTs. When the Internet's internal structure in Syria was analysed, it could be seen that it was closely centralised and controlled by two government agencies, Syria Telecommunication Establishment (STE) and the Syrian Information Organisation (SIO). STE was in possession of all fixed-line infrastructures. Even to open an Internet café, owners would first have to obtain approval from the STE (Freedom House 2012). Moreover, they were expected to control Internet users in the cafés by collecting all names and IDs (Heacock 2011). Thus, they carried out surveillance in the name of the state. Another control mechanism was the SIO, which controlled all gateways through which private ISPs and mobile phone Internet providers connected (Freedom House 2012).

With regard to mobile phones, 3G connections were still expensive in Syria, and not everyone could afford them. In addition to this, both main mobile phone providers Syriatel and MTN belonged to Rami Makhlouf, a cousin of Bashar al-Assad (Reporters without Borders 2012: 38). All of these attempts to centralise ICTs assisted the state's filtration and surveillance endeavours. In addition, the weak and overburdened infrastructure that appeared as a result of the centralised system often led to slow Internet speeds. The state would benefit from these slow speeds during the protest period. Unlike Green Movement protesters of Iran, who broadcast their action to other citizens and the wider world with remarkable speed in 2009 (Shirky 2009), Syrian protesters could not easily upload their videos using their mobile phones. In order to obtain the highest Internet speed, the Syrian activists used dongles. During our interview in Istanbul Abbas,[2] a Syrian activist from Daraa, who later became a member of the Syrian National Coalition and representative for the Revolutionary Movement in the town of Horan, drew attention to the challenges posed by the slow Internet connection in Syria, saying:

> In Turkey, the Internet speed is 5 Mbps. In Syria, my Internet was 1 Mbps and this was the highest speed. So, the Internet was very slow. When we wanted to see the demonstrations, we looked at YouTube and it would say loading, loading, loading... The Jordanian Internet was better. When I went to Damascus, I went to a friend's house. He opened his laptop, put on a dongle and we watched al-Arabia TV. The Internet speed was 5 Mbps. This was the only time that I was able to use the Internet for watching TV.

Despite recent developments to the telecommunication infrastructure, similarly to Abbas most users in Syria were still connecting to the Internet via a dial-up connection and fixed-line telephone subscription. This restricted them to speeds of only 256 Kbps (Freedom House 2012). Although, over time, the protesters obtained higher Internet speed via dongles or satellite Internet, in the first months of the uprising, the slowness of the Internet was an advantage for the government. The state gained privileged access to discourse and communication.

Offline, to appease an increasingly angry revolt, Bashar al-Assad sent a delegation to offer condolences to the families of those killed in the clashes in Daraa during the first week of the protests ("Officers fire on crowds" 2011). The delegation comprised Vice President Farouk Sharaa and Vice Minister of Foreign Affairs Faisal Muqdad. General Rustom Ghazali, one of the highest-ranking members of Syria's Military Intelligence, was also amongst the delegates. The arrival of the delegates brought hope to the people of Daraa, who were expecting an apology from the delegates and asked for the punishment of the officers who killed and injured the people of Daraa (Macleod 2011). Despite these expectations, the state continued to threaten the protesters in its verbal statement. Abbas[3] described what happened in Daraa in these days:

> Rustom Ghazali came to listen to us but he was talking about other things. In the square, there were nearly 100 people. Revolutionary forces, doctors, teachers, engineers... We all went to speak with him but he was threatening us. He said, "You must shut your mouth or you will die". We were expecting that the regime understood the problem and Assad would come to the city to speak with us and apologise. He should have punished those responsible, but what he did was the opposite. Assad sent his soldiers and ordered them to shoot directly into the people. On 23 March (2011), we had almost 60-70 martyrs. We were very surprised by the soldiers' acts.

Abbas's account shows that the inconsistent and arbitrary acts of the delegates who came to Daraa augmented the tension. As Goldstone (2011) writes, repression is likely to end the movement or drive it underground when the state is able to focus its repressive measures squarely on the movement's supporters, but where the repression is unfocused, inconsistent and arbitrary, or where it is limited by international or domestic pressures, it will help the movement attract more supporters (Goldstone 2011: 130). The arbitrary acts of the delegates provoked the mobilisation of protests in different part of the country (Slackman 2011).

Mobilisation Through the Lead of Experienced Activists

Similarly to Egypt, the leading activists in cities took the initiative to organise and mobilise the protests in Syria. As an experienced activist from Darayya, my interviewee Osama[4] was one of these leaders who mobilised the crowds with his friends on 25 March. During our interview in Istanbul, Osama[5] in particular referred to the date of 25 March as the protests spread to several cities, including Homs, Hama, al Tah, Jableh and Latakia and in the towns surrounding Daraa and Damascus on that day. There were also smaller protests in the major parts of Damascus and Aleppo (Free Syrian Translators 2011). Osama mentioned that the protesters took the streets for a better Syria, which belongs to its people and not only to specific sects. They did not want to see the intelligence services everywhere, particularly in schools, where they had more influence than doctors. When I asked what they did in order to mobilise people in Darayya, Osama responded:

> We were concerned about the security forces, so we only used our mobile phones to plan meetings. We did not say anything on the phone. We discussed things face-to-face when we met. We formed circles of five people, then each circle tried to mobilise people. After creating big numbers, we gathered in the mosques. Later on, we went to the street and started chanting our slogans. People who saw us followed us, but the core groups were formed by circles of five people. This was because of security concerns.

Osama's words show that the mosques were the starting point of the protests. Even Christian interviewees of this research mentioned that although they do not pray, they also went to the mosque on Fridays in order to meet with other protesters in Homs.[6] While the mosques operated as the meeting points of protesters, they eliminated the organisational and logistical issues that a social movement organisation would encounter. Moreover, the protesters did not need to put themselves at risk by showing up alone in the protest area; they could cover themselves easily by acting as though they were only there for prayers. The mosques helped people discover that they had not been alone and spurred them to run the risks. They became new protest sites where people gathered and felt deep-rooted feelings of solidarity.

To enlarge their network, Egyptians also gathered in the mosques and tried to activate people who attended Friday prayers. Osama and his friends, drew on the same methods as the Egyptian protesters, created emotional slogans and used these to connect with others. Furthermore, inspired by the Egyptian and Yemeni protesters, the admins of the Syrian Revolution Facebook Page had given a specific name to each Friday.[7] These names were used as motivational factors, to give the protesters courage. Later, when the protests were more systematically organised, the administrators also adopted a democratic approach and let the public choose the name of each Friday, with online polls that were launched every Tuesday night. Yet, Osama's words show that due to the high number of security agents in each segment of society, Syrians also had to adopt unique methods in the offline sphere, such as dividing into small groups and communicating with people who they already knew. Similarly to Osama, Mizyan,[8] another leading activist, also said:

> If you were a group of 20-30 people for the protests, it was good because most people were afraid to go to demonstrations with people they did not know. They might be from the security forces. So, we did this with the people we knew.

The protesters enlarged their group by using the networks of their friends or relatives. As per the theory of the American sociologist Granovetter (1973), they established weak ties, which by meeting the friends of friends would eventually turn into strong ties. Granovetter claims that people who do not have weak ties are unlikely to mobilise in collective action within their community successfully, as weak ties form bridges between people (1973: 25). In order to create the required ties, the core group divided into smaller groups of approximately five people, and each individual in these small groups brought his/her friend to the protests. The members of different small groups concentrated on specific people. This, in turn, enlarged the number of protesters. Social networks of friends and colleagues which established thanks to years of local dissidence had a significant role in this process (Aslan 2015). During our interview, the Turkish reporter Levent,[9] who has worked in Damascus for many years, gave an interesting example to describe how this communication network was developed. She said,

> The best news source was people themselves, you could hear what happened in Aleppo on that day from a taxi driver in Damascus; the communication network was that developed.

Hence, word-of-mouth and snowballing mobilisation tactics were the main repertoire of contention of the Syrian protesters. Due to the fear that reigned in the society, leaders in local areas did not have time to create trust relations with different groups; therefore, they preferred to use this conventional and secure method.

The protesters also utilised old communication platforms, such as coffee shops, for the organisation of the protests. Mizyan Altawil,[10] for instance, as the leading activist from Hama, mentioned that between 18 and 25 March they mostly met in coffee shops during the day and planned the protests in these places. Contrary to what would be expected from a protest organisation in the twenty-first century, where people often reach each other via the Internet without the necessity of meeting, the Syrian protests showed offline platforms were still popular. The use of offline areas and oral communication actually reflected the communication cultures of the Arab world (Rinke and Roder 2011). Oral communication has a striking advantage in creating the resource of trust, which was necessary for collective action. However, the interviews for this study showed that the activists did not only use face-to-face communication because it conforms to their communication culture, but also for security reasons.

Another mobilisation tool was the tape recorders. The power of music had long been discovered by activists and incorporated into the protests to educate, motivate and raise the consciousness of the public. The songs would affect the protesters both emotionally and intellectually and foster group solidarity (Berger 2008). Music had also a unique role to play in the Syrian protests. In order to avoid capture, activists used tape recorders as mobilisation tools. Nadia,[11] for instance, was an Alawite activist from the Alawite-populated city of Tartus. Tartus was a coastal city, known as the city of the Alawite minority, the same Shiite Muslim sect to which Bashar al-Assad belongs (Macfarquhar 2012). In 2011, Nadia was a student in Damascus. As one of the few members of her family that supported the opposition, she was shunned by many relatives who were against her ideas. She thought, as an Alawite, that she was different from them. She was not supporting the state because she was living in Damascus, where she was not only surrounded by Alawite people but also by people from different backgrounds. When the uprising started, Nadia and her friends were influenced by the arrest of 15 children in Daraa and by the repression attempts by the state towards the Daraa citizens who had demanded the children's release. In order to show their solidarity with the people of Daraa, she and her friends decided to ignite protests in Tartus. However, it was challenging to organise protests in this coastal city populated by Alawites. As a main communication method of the protestors in Tartus, Nadia referred to music:

> We spoke with people in Tartus secretly against the regime. Our aim was to represent the Alawites in the demonstrations in Damascus, Homs and Daraa. We tried to fill the streets with balloons and pictures saying that "We are from Tartus", "Tartus is not with the regime". In this way, we tried to help people there. For instance, we gave them tape recorders. They were leaving the tape recorders playing songs of Qashoush on the streets.

Qashoush was a protest singer from Hama. He was famous for singing and writing songs mocking the Syrian President and Ba'ath Party. Playing Qashoush songs in the streets had two advantages. First of all, as the citizens who hid the tape recorders in the streets of Tartus did not need to expose their identities, they were not directly at risk of being subjected to harsh repression. Nadia emphasised that the city did not really need a security service because the society was already acting as a member of Shabiha and that according to Nadia "people could arrest their brother". Similarly to the Alawite actress Fadwa Suleiman,[12] who aimed to raise the Alawites' consciousness, Nadia and her friends also sought to inform people in Tartus that the protesters were not the armed Islamic groups who wanted to overthrow the regime, as was claimed. Knowing that giving political speeches or openly protesting in the city was dangerous, tape recorders became the only secure solution for Nadia and her friends to reach the silent public.

Secondly, the music was the best emotional display likely to gain public attention. As Calhoun claimed, "amongst the things movement organisers need to manage, amongst the tactics they may employ, amongst the strategies they may use against their enemies, emotions and manipulation ought to figure more prominently" (Calhoun 2001: 49). In line with Calhoun's argument, Syrian protesters fought against the cult of Bashar al-Assad via the emotions that the music produced. To show what others thought in Syria and to demonstrate the problems that they were enduring, music was an effective tool. The Qashoush song, "Yalla Irhal Ya Bashar", "Come on Bashar! Leave", was designed to show the enthusiasm of the revolt and ferocity of the crackdown. The infectious refrain of the song was described by the activists as more festive than funny, chanted with the joy of doing something forbidden for so long (Shadid 2011b). The song was taped and widely viewed on YouTube as well which helped to enhance the prominence of the song ("Silencing the voice of freedom" 2011).

The Secret Internet Groups

The cruelty and oppression of the Syrian state also led dissidents to utilise ICTs in unique ways. The experienced activists such as my interviewee, Mazen,[13] who actively tried to mobilise the public on the ground, claimed that:

> We started to create secret groups on Facebook and these groups were called "information groups for demonstrations". They were closed, secret groups. On these pages, we said that there would be a demonstration in front of this bank, on this date, so tries to be there. We took this news and circulated it in trusted circles on the same day. And we got ourselves there, that's how it started.

Hence, rather than launching open Facebook pages through which they could interact with millions, Syrian activists preferred to create closed Facebook and Skype groups. They called these groups "secret Facebook groups" or "secret Skype groups". They interacted with friends or friends of friends in these groups (Aslan 2015). Using social media was risky and dissidents were afraid of being caught by the Shabiha while operating the new technologies. Compared to the well-established Egyptian blogosphere and prominent Twitter users who openly discuss their ideas in Egypt, bloggers in Syria could not even openly announce their support of the people's demands in the country (Ghrer 2013: 116). Syrian blogger and activist Hussein Ghrer (2013) emphasises that although a lot of the bloggers are known for their outspokenness on issues about women's rights and the treatment of disabled people, etc., they would prefer to remain silent and not play a major role in the mobilisation of the protests during the earliest months of the movement. The risk of being caught and the weak repertoire of the activists led Syrians to create closed social media pages. Hence, the Internet was not the main mobilisation tool of the Syrian protests, the Syrian dissidents rather used the Internet to discuss logistical and strategic issues about the protests.

As an alternative method, some experienced activists attempted to organise demonstrations by using their own social media pages rather than creating closed groups. Iyas[14] was one of them. Iyas used to be a tourist guide in Aleppo. When the protests started in March 2011, he returned to his hometown of Saraqib, which had belonged to the Idlib Governorate, and tried to organise a protest in his city in support of the Daraa protesters. He informed people about the protests on his Facebook wall and invited them to make demonstrations with candles in Zamali market. Yet, he described this move as a silly act:

> I was not very smart when I used my own name on Facebook. My friends and my parents warned me about this. They said, "If you want to write, write using another name". But, I wanted to use my own name as I thought a fake name would create problems for me in the future.

Iyas' words actually reveal a dangerous problem with the Syrian protests. Acting on the safe side and operating with fake names created two problems. First of all, it gave way to ambiguity and chaos in the online sphere, a sphere which offered a dangerous illusion of unmediated information flows. Journalists, regular spectators from abroad and Syrians who followed YouTube videos, Twitter accounts or Facebook postings might have thought that they were receiving accurate information. However, as it is difficult to confirm the credibility of an anonymous person, they might easily receive misleading, partial and ulterior-motivated narratives (Lynch et al. 2014: 5). This created disagreement between dissidents who wanted to diffuse peaceful and accurate information and those who wanted to change the narrative in a suitable way for themselves. Secondly, in order to gain the trust of people and activate them, the protesters needed to use their own names. After all, "whether a person trusts a leader is dependent to whether there exist intermediary personal contacts who can, from their own knowledge, assure him that the leader is trustworthy" (Granovetter 1973: 1374). If Iyas had acted anonymously, he would not have been able to rely on his personal contacts or the network that could be formed via these. He would need time to create trust relations with the Internet users. Thus, some of the Syrian activists had to put themselves at risk in order to reach their networks and activate them.

The situation was no different for mobile phone communication. Interviewees for this research said that, from the start of the uprising, they always used secret words when they spoke to friends on the phone. Dima,[15] for instance, one of the women activists from the suburbs of

Damascus, had to flee Syria due to an incautious phone conversation with a friend during a protest. Dima claims her friend was arrested due to their phone conversation, and she had to immediately leave the country. She stated that, when they called people from protest sites, they put this person in danger as well. The Moukhabarat (secret service) was widely listening to the phone calls of certain citizens and could easily arrest them within a few hours.

The statements of my Syrian interviewees show increased public awareness about the phone tracking tactics of the Syrian government. When my interviewees were explaining how they reached their closed circle network via mobile phone, they raised their voice and emphasised the importance of using secret words during mobile conversations about the protests. They also noted that the Syrian protesters used Saudi or Jordanian SIM cards to speak to international media channels. They changed their voices on the phone by using napkins and concealed their identities in order to not be caught by the security services. All of these precautions by the activists, such as using the SIM cards of foreign countries, show that Syrians were well aware of governmental control of ICTs and always assumed that they would be tapped during their phone conversations.

Self-Protective Phase of the State

The first public speech by President Bashar al-Assad since the protests began came on 30 March 2011, twelve days after the spark of the uprising. In this speech, he described the protests as a continued conspiracy against the country and claimed that Syrian society had succeeded in facing it down such challenges every time, which enhanced its strength and pride (Landis 2011a). If his speech is analysed employing a critical discourse analysis, it can be seen that, like other Arab Spring leaders, Bashar al-Assad adopted a nationalist rhetoric. He claimed that the country was at war against foreign powers that undermined Syria's stability. Thus, he adopted a discourse of "we" and "them" and paid extensive attention to the alleged threats of foreign powers to the interests and privileges of Syrian society. Such a strategy is conducive to the formation of themes that take a "we are the victims" stance (Van Dijk 1993: 265). This could have led to the formation of a negative attitude towards the protesters. Bashar al-Assad also adopted Pan-Arabist rhetoric and emphasised the importance of Syria for Arab unity and Pan-Arab rights. The president claimed that "they" prefer "Syria to get weaker and disintegrate because this will remove the last obstacle to Israel's plan" (Landis 2011a). The discourse of Assad reflected the political culture in Syria, which was alarmist. Referring to Israel and the West, Assad was trying to keep the fear of instability alive in the society.

In most parts of the speech, biographical narratives were also apparent. Berenskoetter (2012) defines biographical narratives as narratives by which communities define themselves in time and space. For instance, states often reflect a biography of their history, character and aspirations. In his speech, while Bashar described himself as "the faithful brother and comrade who will walk with his people and homeland", he projected the state as internally and externally different from other Arab countries, open to reforms and faithful to Pan-Arab rights (Bashar al-Assad, March 2011). Syria was not a copy of other countries. It had seen development; an opening up process and communication between he and the Syrian people had already existed. By emphasising the difference between Syria and other Arab countries, Bashar al-Assad wanted to persuade the public that there would be a change, but one different to that in other Arab countries, a change *under* his leadership.

When he talked about new technologies, like other Arab leaders and Turkish Prime Minister Erdogan, Bashar al-Assad chose to demonise satellite TV stations and the Internet. In her article on Turkey's Twitter ban, Zeynep Tufekci argued that what the Prime Minister of Turkey wanted to achieve was to demonise social media channels by placing them outside the sacred sphere, as a disruption to the family, as a threat to unity, as an outside blade tearing at the fabric of society (Tufekci 2014b). In his speech, Bashar al-Assad also aimed to demonise these new technologies by claiming them to be foreign countries' and extremists' tools, which assisted in their acts of sabotage. He asserted that "It is a virtual war and they are waging on us a virtual war using the media and Internet". Social psychologists claim that how people process, store and apply information about certain social events can shape their understanding of these events (Dijk 1993: 257). In accordance with this theory, Bashar al-Assad attempted to create a new social cognition about the uprising and protesters by painting the protests as being dangerous to national unity.

On the ground, the state felt it is necessary to organise pro-Assad rallies in order to show the popularity of the leader to the world. On 29 March 2011, just before the first public speech of the president since the protests began, these rallies took place in Damascus, Aleppo, Homs,

Hama, Tartus and other cities (Lesch 2012: 95). The counter-protests were also helpful to show that certain segments of the society still supported the state and that unity amongst protesters and the public had not yet been achieved. In order to protect the image of the president in the eyes of the public and the "cult" on which his power rests, another tactic was to flood the public space with his iconography (Index on Censorship 2011).

Co-ordination Through the Lead of Local Committees

President Assad's first public speech created disappointment amongst the public, and on Friday, 1 April 2011, the protests grew further and appeared in Daraa, Latakia, Baniyas, Homs and Qamishli, as well as smaller settlements such as Amouda, Tel Tamer and Ras Al-Ayn (Marsh 2011b). Over the following months, the protests took place each Friday across Syria with the assistance of an increasing number of protesters. In some cities, protesters took to the streets every single day. The interviews for this research showed that although the activists of the same city were in contact with each other via the mobile phones, they were not in co-ordination with the activists of other cities. However, the fatal clashes between the protesters and police indicated that for a successful movement, the protests needed to move beyond spontaneous and unco-ordinated action towards a more organised campaign in all Syria (Najm 2011). In pursuit of this purpose, the groups that were originally established to organise and document local revolutionary activities on the ground were unified at the Local Coordination Committee (LCC) level (Aslan 2015). The first committee emerged in Darayya, a restless suburb of Damascus (Shadid 2011a). As a leading dissident in the Darayya protests, Osama[16] described the merger of the LCC and Darayya local committees:

> The Darayya Local Committee was effective in organising protests and other peaceful activities. Actually, it was founded by a number of leading activists in the city before the foundation of the Local Coordination Committees in Syria. When the LLC was established, the Darayya committee chose to join it and remained active within it. (Aslan 2015)

Hence, some leading figures in towns and neighbourhoods, such as Osama, became pioneers of the oppositional unity. As a result of these efforts, the first organised committee in Syria emerged from Syria's restive streets.

Even though the committee in Darayya took its power from the street, the social networking platforms loomed large in the mobilisation and internationalisation endeavours of the LCCs' members that organised across different regions via this tool (Shadid 2011a). Nearly 120 local committees emerged across the country and the representatives of these groups started to meet every two weeks via Skype to co-ordinate and synchronise activities. LCCs became responsible for reporting news and developments on the ground to international media by gathering, checking and providing real-time information about the uprising (Aslan 2015). They also constantly updated the LCC website[17] and Facebook page (Local Coordination Committees of Syria 2012).

The interactive networks in different cities were also an important step for the creation of a collective identity and sense of solidarity in the protests. Mazen,[18] who was helping the Homs Local Committee Council, talked about the attempted transition from spontaneous to a more organised form of protest:

> You meet people in the protests and start to create your own trust circle. Then, you start to choose different places to meet with them outside the protests. This was how the local co-ordination committees were established. Actually, where we were meeting was in the Internet café of a friend. Aside from organising the protests, we were creating logos and banners, drawing some flags, preparing the sprays. This was how the local co-ordination worked. We also wanted to know what was going on in other cities and how they organised. Thus, we started to create Skype groups. In Skype groups, we created co-ordination groups. For example, we contacted local groups in Daraa. They told us to use onions when the gas came or not to organise the protests from the same mosque. From then on we changed the mosque every Friday.

The words of Mazen demonstrate the improved co-ordination amongst different local groups on the ground. The co-ordination in the cities helped to form solidarity networks that had never existed before. For instance, it helped in the formation of solidarity between cities with historic rivalries, like Homs and Hama (Shadid 2011a).

The connection also had vital importance for the safety of protesters. For instance, as a software engineer living in Hama, Mizyan[19] had to leave his city and went to Daraa to fulfil his mandatory military service. During our interview, he stated that in the army, where mobile phones and the Internet were forbidden, he secretly checked his mobile phone and mostly used Skype to communicate with people and warn them about the army's activities. He trusted Skype communication more than mobile phone services, as security tools such as TOR or HTTPS provided a safe Internet connection to the protesters. He specifically emphasised the role of secret Skype groups in this process. He informed protesters in cities about the arrival of army tanks through these groups. While mobile phone use was tightly monitored by the state, Skype groups of the LCCs emerged as the main communication method.

Internationalisation Attempts of the Activists

In the beginning of the protests, the low speed of the Internet restricted the framing capabilities of dissidents and made them and the public dependent on traditional media channels or news from diaspora. As a result, a reciprocal relation appeared between the international media channels and the protesters. While outside media had to use the protesters in order to gain access to those on the ground and get information, the protesters could reach the rest of Syrian society via the international media channels (Aday 2012). The diaspora also took an active role in the accessibility of news to the outside world. Initially, the diaspora constituted three networks (Almqvist 2013: 52). The first of these was comprised of various exiled, well-educated and cosmopolitan intellectuals, such as Radwan Ziadeh, Ammar Abdulhamid and Ausama Monajed. The violent crackdown of the state on opposition and civil society movements during the Damascus winter forced these activists and dissidents into exile. The second network had been formed by younger, first-generation immigrants of Syrian descent. They were mostly the children of political exiles who had fled the country during the turbulent 1980s or later. Finally, the traditional opposition in exile including the Muslim Brotherhood, Kurdish nationalist tribal leaders and various leftist parties formed the third diaspora group (Almqvist 2013: 52). These three groups, and particularly the young and Internet-savvy living abroad, started to co-operate and brought help to the protesters from the beginning of the uprising. Abed,[20] for instance, was a Syrian activist who lived in Jordan his whole life. When the protests erupted in Daraa, he immediately helped the dissidents in Syria by uploading videos on the Internet and writing reports to the media agencies. During our interview in Istanbul, he described the active role of the diaspora by saying:

> Some people from Ar Ramtha city in Jordan had relatives in Daraa. I had friends who used to go there. They started to sneak out flash drives full of videos. We started to bring them to Jordan and uploaded them because the people of Daraa could not upload anything. So, at some point, the Internet stopped in Daraa. The activists had to go to Damascus just to upload the videos or smuggle them into Jordan.

Abed was actually referring to the first days of the uprising. During the massive protests in Daraa on 18 March 2011, buildings were set on fire and vandalised. Troops surrounded the city and the phone and Internet connections were interrupted (Marsh 2011a). As the Internet was very slow or completely gone in many areas, the activists had to rely on their contacts outside the country.[21] The spontaneous grass-roots videos filmed by unknown protesters in Syria and sent to the diaspora would then facilitate the dissemination of information about the protesters. The diaspora multiplied and expanded the visibility of these crowds and supplemented the efforts of activists inside Syria by connecting these groups to Syrians abroad.

As the diaspora activists developed their repertoires of contention, their ICT use and media reporting also became well-structured and better prepared. For instance, the Syrian journalist and activist Amer,[22] who was based in Jordan when the protests started, explained this process:

> We started doing new hashtags. For example, if we had something for Damascus or a massacre in Damascus, we created special hashtags for Damascus. After that, we thought of doing something special. We had, at that time, lots of contacts with people who had Facebook pages and connections to the media. So, we started to do online posters and awareness videos. At the beginning, we were maybe 100 or 150 Syrians tweeting about Syria; we were very few but now there are maybe 1000 people.

I conducted this interview with Amer in 2013. His words emphasised the gradual improvement in the online activities of the diaspora activists.

Due to the absence of journalists and professional cameramen, it was also difficult to find professional documentation of the massacres committed in Syria in the first months of the uprising (Local Coordination Committees 2011: 5). My interviewee Hani,[23] who was an activist from Damascus, described the absence of documenting as the biggest problem of the Syrian uprising. He said:

> We had a problem in documentation of the massacres. This was the main problem in our revolution. The US, as well as the human right organisations, did not pay attention to the un-documented massacres and we couldn't prove them. I think it is important today to learn how to document. This footage will remain after Assad. If we want to put Assad in prison, we will need these videos.

Hani pointed out an important problem for Syrian dissidents, namely that the only way to prove the accuracy of the news was to document and record the atrocities committed during the uprising (Lynch et al. 2014: 13). Undertaking the documentation initiative was difficult for regular citizens, as it brought different risks with it. The emotions of fear, mistrust and intimidation that related to weak repertoire in the country might also be the reasons that hindered such initiatives (Naidu 2013: 4). As the dissidents made solid connections in the offline and online spaces, they also became more courageous in taking initiatives.[24] The act of filming would take centre stage in the dissidents' activities, and they would create their own tropes, formats and genres for documenting political activism in Syria's heavily surveyed public sphere (Della Ratta 2018).

The citizens' media culture has also gradually spread and encouraged creative and collaborative work in the online sphere (Ghrer 2013: 119). While the number of citizen journalists increased in the online sphere, the number of Syrian-initiated news websites and online radio stations increased significantly. Harkin (2013: 102), who analysed newspapers and news websites through a small-scale ethnography and informal interviews, cite the journals *Hurriyat*, *Al Haq* and *Souriatna* as the new revolutionary newspapers, which started to post on YouTube and Facebook. The interviewees of this research also indicated Souria Houria[25] as the best news website reporting on Syria with its translations of the news into English, French, German, Dutch and Spanish. Rabee,[26] an engineer living in France, is the technical administrator and editor of this news website. When asked why they opted to open a news website rather than operating on Facebook, he said:

> All our accounts on Facebook and Twitter are linked, but we choose to have our website so we have our database and our own space; we can keep documents, videos and archives. Our country is living a very critical transition and we need to follow and document all events. The future of a new

democratic Syria needs this data and information, as you know Facebook has deleted a lot of accounts and pages in the past, and so we wanted to have our own space and be the master of it.

The restriction problem that Rabee mentioned with regard to Facebook was a significant problem that the protesters still face today in social networks. As admin of the Syrian Revolution 2011 Facebook page, Mahmoud[27] also complained about the restrictions they faced on Facebook. He spoke of their desire to open a new platform on which they can determine the rules themselves. In 2013, Anonymous also called on Facebook users to protest against the social network's alleged censorship (Merrett 2013). The Syrian protesters, who do not want to be restricted by Facebook rules, launched their own websites and tried to draw the attention of the international community to their news.

Nevertheless, despite its strict rules and restrictions, Facebook has still been a popular platform for Syrian protesters. In addition to the Syrian Revolution 2011 Facebook page, many new Facebook pages have also sprung up since the uprising began, the most prominent being: the Syrian Days of Range,[28] the Shaam News Network,[29] With You Syria[30] and We are all Martyr Hamza Alkhateeb[31] (الخطيب علي حمزة الطفل الشهيد كلنا) (Almqvist 2013: 57, Interview with Sama). These e-magazines, opposition web pages and social media groups became the voice of the revolutionaries and their tools for shaping the narrative of the protests according to their identity and policy narratives (Harkin 2013: 102). Sama,[32] for instance, was one of the admins of the "We are all Martyr Hamza Alkhateeb" page. When I asked her why she became the administrator of the page, she stated that in the absence of social networks, they would all be killed without anyone noticing. She added:

> When I saw how they tortured this innocent child who was trying to deliver milk and bread for the children of the besieged city of Daraa, I thought it was our duty to expose this crime to the world.

The main aim of Sama was to demolish the identity narratives of the Syrian state that described the state as reformist and that it served its people. She tried to get the support of the international community by showing the torture and other atrocities that were ongoing in Syria. Sama and other activists would use these platforms as substitutes for a space violated and dominated by the regime. As Della Ratta (2018) argues, the videos shared on the social media platforms would make the activists visible and render them the essential place to perform and manifest dissent.

An increasing number of media centres have also been launched since the start of the protests. The Activists News Association (ANA) Cairo and the Syrian National Coalition Media Centre (SNCM) were examples of Syrian dissidents' efforts to shape the news in the international arena and to gather the support of the international community. The ANA Cairo Media Centre was located in Cairo. It distributed easy-to-use flip cameras to activists inside Syria in order to cover events in the country (Almqvist 2013: 60). The SNCM was another initiative of the opposition to follow and distribute news about the uprising. Directly linked to the opposition outside Syria, the centre provided daily reports and news about the Syrian uprising and released press statements. Based in Istanbul, the Media Centre also developed and conducted international training workshops inside Syria. The employees of the centre told me that they were in contact with the activists inside Syria through the satellite Internet that was provided to those activists by the centre.[33] Citizen journalism inside Syria flourished and developed with the help of these centres, which competed to shape the narrative for the outside world.

Hence, Syrians in exile and the activists inside had closely co-operated in order to disseminate information coming from the Syrian dissidents and gather support from international society. It could be said that, to some extent, the co-ordination of the protesters with the international media broke the media blackout of the state. As Volkmer (2014) states, it also gives way to the process of de-bracketing the society-state nexus. Volkmer (2014) defines the de-bracketing process as fracturing the public communication characterised by the transformation of publics, from national mass media spheres to the complex structures of interactive transnational networks (Volkmer 2014: 34).

Strategic Manoeuvres of the State

In April 2011, the United Nations declared that at least 1100 people had died in the Syrian uprising since the start of protests in March ("Syrian crisis" 2011). While the state continued to crack down on demonstrations, it started to introduce reforms in the country as a public strategy. Citizenship was granted to the 220,000 Kurdish people who had been living in Hasakah ("Stateless Kurds in Syria" 2011). To calm conservative Muslims, the state removed a "niqab ban" in schools only introduced the previous summer ("Syria relaxes veil ban" 2011). Other appeasement attempts included the release of prisoners detained for a couple of weeks and the appointment of a new Prime Minister, although the whole country knew that the Prime Minister possessed no genuine power (Ajami 2012: 75). Thus, after the "Damascus Spring" in the 2000s, Bashar al-Assad introduced a reform period for the second time.

The second speech of Bashar al-Assad was also based on this flurry of reforms (Landis 2011b). The speech came just after the swearing in of the new government on 16 April. In comparison with his previous speech, it could be said that this one was more tender and reconciliatory, mostly emphasising the problems in Syria that needed to be addressed by the state. The president, who addressed ministers, admitted the necessity of introducing reforms in the country. He also acknowledged that he had been absent from the public eye for some time, but was keen to bring the changes that the public wanted (Athamneh and Sayej 2013: 174). As one of the most important changes, he noted the need to shake off the corruption in government once and for all. He drew attention to the emergency law and emphasised that the new ministries should work hard to lift the state of emergency as soon as possible.

This second speech of Bashar al-Assad projected an identity narrative through representing the state as reformist and at the public's service. Despite these reformist overtures, when he described the protests that were going on across the country, the president used the same discourse of "we" and "them". He referred to the protests as a conspiracy directed against Syria by foreign powers. The discourse adopted by the president was also reproduced and supported by state TV. While the national TV channels mostly gave space to pro-Assad rallies in their coverage, they framed images of the opposition during protests as attacking soldiers ("Syria protests: making amends" 2011).

The interviewees defined the repetition and reproduction of these messages as an intelligent act that helped the president control knowledge in the country and produced cynicism amongst the silent public towards the activists. Coming from a Christian background, Caroline,[34] for instance, stated that:

> From day one, they said: "Those are Islamist groups", "This is a war on terrorism" and "It is a worldwide conspiracy against Syria". The regime has played this tune very well from the very beginning, and that's why Christians were afraid. When people armed themselves, some Christians said: "you see, it is an armed rebellion". They weren't even calling it a revolution. Mainly, Christians from my community were very scared.

By describing the protests as a conspiracy of foreign powers, Bashar al-Assad created the perception of an enemy against which the nation should unify and resist (Athamneh and Sayej 2013: 175). It was easier for the state to create a dominant narrative and shape public perception in Syria, where the national media was a vocal proponent of the state and foreign reporters were banned from entering the country.

On 21 June 2011, the same discourse was used by Bashar al-Assad at Damascus University (Landis 2011c). Assad divided the opposition into three segments. He identified the first of these as the people who had demands and needs, the second as the small group of people who manipulated others and were wanted for various criminal cases and, finally, the third as the people who held extremist and takfiri ideologies. Like the national media, the president claimed these last two groups to be the people behind the crimes committed in Syria. Assad said, "In some cases, peaceful demonstrations were used as a pretext under which armed men took cover, in other cases they attacked civilians, policemen and soldiers by attacking military sites and positions or used assassinations" (Bashar al-Assad 2011). Here, the president once more adopted the discourse of "we are the real victim" and blamed the opposition for the chaos in the city. Most interestingly, in this speech, Assad chose sectarian strife as his emergency plan for survival. He tried to destroy the alliance between different sects and produce a sectarian conflict by criminalising opposition activities and creating cynicism amongst the people. At some point, these tactics helped the state and mistrust between religious groups increased (Almqvist 2013: 34). This discourse was also important to keep up the fragile alliance between religious minorities and the moderate Sunni merchant class. After all, Sunni merchants had no desire for instability. The president emphasised that, in his absence, Syria would become a state of rival tribes and, in order to avoid this division, he asked the public to keep the "great spirit and deep sense of identity" (Bashar al-Assad 2011). Such nationalist rhetoric often contains negative statements and decisions about minorities (Van Dijk 1993: 266). In this case, protesters were described as minorities against whom the whole nation should unify to protect their homeland.

Yet, these narratives were not sufficient to pacify the crowds that had already taken to the streets. In order to derail the movement completely, troops and tanks stormed Daraa on 25 April 2011. The Army's occupation of Daraa continued until 5 May (Holliday 2011: 13). Abbas,[35] who fled to a village near Daraa, described that day:

> The morning of the April 25 (2011) will not be forgotten. I was in the village because the regime was after me and I was hiding there. We woke up in the morning; electricity, telephone and the Internet were all off. In an hour, we learnt that the regime had attacked Daraa and there had been no information about Daraa for five days, no information at all. Everything was off. When we tried to enter the city [Daraa], soldiers forced us to return to the village. One of our friends in the city had a satellite phone and was speaking to the TV but he could not give enough information on what was going on in the city, as he was in his house. He only described what he saw on his street.

The experiences of Abbas show that Syrian dissidents had to rely on alternative communication mediums during times of emergency. Satellite phones were one of them. The difference of satellite phones from mobile phones is that one could call or transfer data from anywhere, as satellite phones entirely bypass the local system (Tobias 2013). This technology became the only means of remote communication for the captive citizens of Daraa. However, the high number of soldiers in the streets prevented citizens from reporting what was going on in other parts of the city. To tighten the information blackout, the government took more precautions and declared that people found with satellite phones would face 11-year jail sentences (Smith 2011).

During the month of May 2011, the state applied the same siege tactics to cities like Baniyas and Homs. Despite the siege, Homs protesters continued to take to the streets and the tension rose even further. Riots subsequently proved that the demonstrators would not give up the protests easily. In June, the enthusiasm of Homs protesters spread to Hama as well. Hama protesters had been active since the start of the uprising in March, but Hama was not seen as a significant site of protests until June. On Friday, 1 June 2011, one of the largest protests, calling for the removal of President Assad, occurred in the city. Two days later, the security forces, mostly been absent in Hama until that time, surrounded the city and killed at least 60 people. In the following weeks, Hama residents (like the Homs protesters) regularly participated in the protests ("Syria: Shootings, arrests follow Hama protest" 2011). In parallel to the repression by the state, the protests in Syria increased; they were seen in several suburbs of Damascus, and in the Midan neighbourhood, in the centre of Damascus. The state intervened in Hama in the beginning of July and used similar tactics as had been applied in Daraa, Homs and Baniyas. Tanks surrounded the city, electricity was cut off and Shabiha began a campaign to arrest local residents and human rights activists ("Syria: Shootings, arrests follow Hama protest" 2011). Meanwhile, the protest singer Ibrahim Qashoush was found dead in a river in Hama on 4 July. The state's response to his songs showed its intolerance towards the protests.

The operations of the state in besieged cities demonstrated that it did not apply the same Internet tactics as other Arab countries. Instead of cutting the Internet entirely, which would have caused significant harm to the country's economy, the Syrian government applied local cuts to cities where the protests had intensified (Eagleman 2012). Cutting protesters' communication to the outside world in particular cities ensured that the lives of citizens in other areas could continue unaffected.

In addition to the state's adaptive Internet strategies, Syria became the first Arab country that had an electronic "army" openly launch cyber-attacks on its enemies. Although there was no concrete evidence linking the Syrian Electronic Army (SEA) to the Syrian state, one could easily see the government's support of the initiative. Despite the tight control of the state over the Internet, the army could operate with impunity over Syrian networks (Brewster 2018). The army's website "sea.sy/index" was registered on 5 May 2011 by the Syrian Computer Society, an organisation that had previously been headed by Bashar al-Assad himself (Syrian Electronic Army 2011; Fowler 2013).

Who were the members of this "army"? The army first emerged on Facebook in April 2011, but its first Facebook page was removed by Facebook shortly after it was created. Since then, the army has established new Facebook pages every time they are blocked by Facebook. After the Facebook account, the army also launched Twitter and YouTube accounts and published its own web page in May

(Noman 2011). On its website, the army members described themselves as a group of young people who did not have any affiliation with the government but were willing to protect their homeland. They defined their aim as fighting back electronically and publishing the truth against the Arab and Western media that showed its bias in supporting the terrorist groups.[36] Since its emergence, the army has attacked opposition groups, government institutions and western media outlets. It took responsibility for hacking news websites such as the Association Press, the BBC, National Public Radio and CBS (Fowler 2013). The army's attacks marked a major offensive by the Syrian government in the online arena. Moreover, the discourse of the army on terrorists and the bias of the Arab and Western media also reinforced government propaganda. As Tufekci argues "in the networked public sphere, the goal of the powerful often is not to convince people of the truth of a particular narrative or to block a particular piece of information from getting out, but to produce resignation, cynicism, and a sense of disempowerment among the people" (Tufekci 2017: 228). The messages of the army fostered confusion, fear and doubt amongst the silent segment regarding the activists. By destroying the factual and credible news, the army also served to foment confusion and makes the activists vulnerable targets.

Another surveillance method of the state was to force detainees to open their social media accounts so police could easily reach other protesters. Bassel,[37] for instance, a Syrian dissident from a suburb of Damascus and the editor of Syrian National Coalition media, mentioned that when his cousin was arrested, they put him in a room and the officer next to him asked four questions. The questions were: "Do you have a computer?", "Do you have internet access?", "Do you have a Facebook account?" and "Do you visit revolution pages?" Bassel added that there was no need to lie to the officers, as they had already monitored these social accounts before arresting people. Bassel words have also been justified by the international activist groups. The reflets.info website, working with the Telecomix digital activist group and fhimt.com, a Tunisian portal, revealed that the Syrian government had been using Blue Coat tools (American-made devices) for monitoring the Internet activities of its citizens (Reporters without Borders 2012). While these tactical agilities of the state would semen fear amongst the activists, they also dispersed the protests' energy.

SUSTAINING PEACEFUL PROTESTS

The first- and second-hand data I gathered for this research pointed out four factors that prevented the sustainability of the protests in Syria.

The Lack of Unity

In the first month of the peaceful demonstrations, while the protests intensified in all these cities and towns, there were no significant anti-regime protests in the two largest cities of Syria, Aleppo and Damascus (Lesch 2012: 95). Being aware that a lack support from Damascus and Aleppo would create damage for the Syrian revolution, Daraa activists tried to organise the next rallies in Damascus on 25 March. Abbas[38] described this process:

> The next Friday [March 25, 2011], the demonstrations were bigger. The villages and towns around Daraa city, as well as the city itself, went out to demonstrate for the blood of our martyrs. We wanted to make our voices heard by the world. We were the citizens of Daraa and we were getting killed by the regime, which was getting even more violent. We decided to change our policies. This regime would not be affected if we continued the demonstrations in Daraa. We decided to change the place of demonstrations to the centre of Damascus. We sent our friends to Damascus by bus each day, about 100-150 people in total. We rented a number of flats in Damascus and sent these people to these flats. We told them, "on Friday, we are going to carry out a huge demonstration in different parts of Damascus".

Despite all these plans by Abbas and his friends, the protests did not spread to Damascus, the capital of Syria and second largest city after Aleppo. One reason for this failure was the significant military presence in Damascus and Aleppo (Lesch 2012: 96). The military was deployed at specific points and was preventing the entrance of protesters from different areas to the cities by checking their identity cards. Abbas and his friends, for instance, were sent back to their cities after their identity cards were checked by the military. In this way, the military was successful in preventing the big rallies that were simmering in the suburbs and outskirts from reaching these cities. According to the Middle East scholar David W. Lesch (2012), another reason for the relative lack of protests in the big cities was the active presence and proximity of Bashar al-Assad. Assad was working in Damascus and travelled to Aleppo quite often. Moreover, Damascus and Aleppo received lion's shares of foreign investment, which yielded infrastructural improvement and beatification. The middle and upper classes of these cities had benefitted from this investment and did not want to lose the economic, social and political benefits they had already acquired (Lesch 2012: 96). The heightened level of violence, the lack of reliable information and the media presence of the state's discourse about the sectarian motivation of the opposition further created confusion and anxiety amongst many in Damascus, including Sunni merchants, the state-employed members of the middle class and old Damascene families (Crisis Group 2011; Lesch 2012: 96).

The Lack of Political Experience

In the first month of the uprising, despite the efforts of leading activists to create a sustainable movement, when the rapid mobilisation attempts of the protesters met with harsh repression from the security services, the protest groups had to dissolve. Mizyan,[39] for instance, mentioned that the first protest attempt in Hama had to end quickly due to the significant presence of security near the protest place. He said:

> So, the demonstration was at the Omar Mosque after Friday prayer. I was observing the mosque from the next street [Mizyan helped to organise this protest but did not join it, as he was afraid to be arrested for a second time by the security services]. People got together here in the mosque and tried to go to the square. But, after twenty metres, security personnel arrested all of them.

Mizyan's words showed the inexperience of the Syrian activists when it came to protests. Like the April 6 movement's protest attempts in Cairo in 2008, the Syrian protesters poured onto the streets without considering and planning how to deal with the police beforehand. When their limited collective action capabilities met with intense repression attempts by the state, their frustration and anger rapidly increased. As the protests increased and intensified, the government's response became more violent (Lesch 2012: 94). Deadly clashes between police and protesters show that the lack of experience is an essential problem that crowd-enabled connective action face today. As Tufekci (cited in Zuckerman 2011) argues, connective action helped dissidents carry out the spectacular street protests, but the dissidents cannot build a larger movement to topple or challenge a government unless the movement entrenches its capacities and strategies before coming to public attention.

In the further stages of the uprising, the LCCs played an important role in creating strategies to follow on the ground. However, these were not sufficient in themselves to form a collective identity amongst the protesters such as the one that had been seen in Tahrir Square during the 2011 Egyptian protests. The most important reason behind their failure was that only 100 to 200 members were fully engaged with the LCCs and these were mostly young, technologically savvy members and many of these people did not take part in the protests themselves, as they were afraid of being arrested. They mostly spent their time in front of computers (Shadid 2011a). On the other hand, in some cities like Darayya, members of the LCCs were the leading activists who had established solid connections to people on the ground and had already been familiar to the leaders of the masses.[40] Due to the variation in the LCC members' activities, the committees charted different directions in different cities (Shadid 2011a). For instance, while the LCCs were effectively active in the organisation of the protests in Darayya, their influence on protest organisation in Homs was limited and they were mostly active in the area of humanitarian aid (Interviews with Mazen and Osama, 2013 and 2014).

The limited achievements of the LCCs also related to the partial capacities of the digitally supported protesters to establish political objectives. Apart from the famous slogan, "the people want to overthrow the regime", the few in LCCs articulated what would come next (O'Bagy 2012: 20). According to Mahmoud,[41] the admin of the Facebook Revolution page, the limited political experience of opposition members was an important factor that weakened the local co-ordination committees and the opposition groups that appeared later. Mahmoud named the Syrian Revolution General Commission (SRGC) as an example that was founded in the early stages of the uprising. He mentioned that due to the fights amongst the leaders of SRGC, the group have had difficulty sustaining themselves. He added:

We have to take into consideration that, in the last 40 or even 50 years, in Syria, since the Emergency Law was imposed in 1957, the Syrian people had no chance to experience political life. Now, Syrians are trying to obtain the experience of 50 years in two or three years.

Mahmoud points out that the damage of the culture of obedience in Syria curtailed the political capacities of the opposition for a decade. The grass-roots opposition, who lacked political experience, failed to launch an initiative that harmonised the efforts of political forces (Al Abed 2013). Hence, the opposition struggled to provide an alternative to the regime and failed to respond to what would come next (Hassan 2011).

The Cycle of Violence

Inspired by the Egyptian protesters, the protesters in Homs and Hama tried a new method to sustain the peaceful protests two months after the protests began. On 18 April, the protesters occupied the Clock Square in the city ("Syria protests" 2011). The attempt of Homs protesters was not successful due to the brutal response from the state. When the protesters occupied the square and commenced the sit-in, the police responded with heavy gunfire. Mazen,[42] who was at this sit-in, said:

> Our aim was to organise a sit-in like they did in Tahrir Square and actually we did something like that on 18 April [2011]. We organised a sit-in at Clock Square in Homs. There were more than 20,000 people. We thought that we had the atmosphere like Tahrir Square and that the regime would go away in a week. We were so happy organising this. After midnight, a lot of people went home. So, we decided that some people should stay and protect this place. But, later, the army came and killed almost 200 people. We did not do something like that again.

The high expectation of Mazen from the sit-in shows how media images associating Syria with the Tunisian and the Egyptian protesters created false hope for the Syrian people. References to Tahrir fixed the image of revolution in people's minds.

The protester in Hama also started to occupy the Al-Assi Square in the city every Friday after the death of 60 protesters during the protests in Hama on 3 June (Interview with Mizyan, 2011). The biggest protest on the square happened on 1 July. Once more the security forces responded to the protests by opening fire, killing at least 16 people and arresting the local residents and human right activists ("Syria: Shootings, arrests follow Hama protest" 2011). The brutality of the state obstructed this process as well and yielded different results in Syria in comparison with the Tunisian and Egyptian protests.

For understanding these cycles of violence in Syria, one can look at Holliday's (2011: 16) work, which provides a superb framework regarding this cycle and reports that although the brutal repression of the state silenced Daraa and Hama, the result was different in Homs. As the conflict dragged on, the protesters resorted to armed resistance in Homs. What happened in Homs demonstrated that the brutal repression had two impacts. It helped to suppress the protests, but also produced an escalation in the violence.

Restricted Access

Despite the violent direction of the protests in June 2011, most protesters were still trying to attack the state in a peaceful way that June and July. However, in the absence of international media access to Syria, digital media could not provide the same opportunities to each Syrian who wanted to be heard by the outside world. In August 2011, for instance, women's groups started to be active in Damascus. Bayan Aladawi[43] and her friends were one group. They mobilised approximately 25 people together for the protests; however, they could not get any media coverage. In order to show their protest video to the outside world, these women activists used YouTube[44] and received 830 views. Compared to the videos of the Tahrir Square protests in Egypt, which got many millions of YouTube views, the number of views for the women's protests in Syria was disappointing. One of the reasons for this difference might be the absence of international media in Syria. In Egypt, a large number of foreign correspondents and human right organisations were ready to document the protests in Tahrir Square; in Syria, YouTube became the primary source of video content for the press and TV channels. However, although the Syrians worked hard to document the protests for international media, the videos could not attract the same attention as those from Egypt. When compared to the post-protest videos that gained more visibility—such as the shelling of Homs[45] (30,056 views), the lung-eating rebel commander[46] (25,243 views)—the numbers of people who watched the first women's protests in Damascus via YouTube numbered only 835. In the former videos, the shocking and violent scenes caught the attention of the mainstream media. This helped in the diffusion of the recordings worldwide. The women's protests could not evoke the same emotions of horror, surprise, violence or joy, which usually make footage newsworthy. One reasons for this failure was that Syrians had to be prudent when they engaged in their first protests. To protect the identities of the women, the footage showed the women activists with their backs turned. Hence, although YouTube was one of the biggest opportunities for Syrian activists, not everyone could gain YouTube-mediated visibility. Bassel[47] was another activist who was stricken by this problem. After the protests started, he became a member of the Syrian non-violation movement. As an experienced activist, Bassel knew that, if the revolution became a violent uprising, those most affected would be the Syrian citizens themselves. To prevent this, he and his organisation put all their efforts into informing people about the importance of peaceful protests. They posted YouTube videos[48] explaining the value of non-violent protests for the success of collective action. Bassel claimed:

> I joined the "Syrian non-violence movement". I was trying to provide activists with the idea and information on how non-violence works rather than arms. I tried to tell people what would happen if they got arms and why it would never work. But, we could not do anything useful.

When Bassel was asked why he thought their attempts did not achieve their aim, he pointed out the number of viewers who watched their YouTube video: 267 views, obviously disappointing for the dissidents. Getting the attention of people and the mainstream media for an educative video was a difficult mission for the dissident groups. The mainstream media was not very helpful in this process. As ABC's Lara Setrakian noted, news media is usually inclined to show what is urgent and this can come at the expense of what is important (Lynch et al. 2014: 11). Since the dissidents who fought for peaceful demonstrations could not produce novel stories or images, they could not get international media attention and lost their chance to frame the news over time. This was seen by protesters as an underrepresentation of certain segments of the revolution.

Conclusion

Tracing the development of initially peaceful protests in Syria shows that, in each stage of the uprising, specific actors and media came to prominence. In the first stage of the protests, the weak repertoire in Syria was effective in shaping the activities of protesters. The organisation of protests could be better understood by examining the Friday rallies. The findings indicate relying on their past protest performances, leading activists in different cities played an active role in mobilising the public. Snowballing mobilisation tactics, wherein small mobilisation groups galvanised their friends or relatives, were the protesters' main repertoire of contention. Mosques also had a central role in facilitating gatherings and creating a sense of "we-ness" in the public sphere. Analysis of protestors' communication methods reveal that word-of-mouth and face-to-face communication were significant communication methods used by protesters. Interpersonal communication in secret Facebook and Skype groups also contributed to the exchange of information and informing the world about the situation on the ground, but the Internet could not be the main organisational hub of the protesters.

The co-ordination of local committees and their internationalisation attempts support the hypothesis of this book, which asserts a symbiotic relation between repertoire of contention and the use of ICTs. As protesters developed themselves in protest fields and form social ties with different groups, their spontaneous activities increasingly became more organised. Local committees in towns and neighbourhoods began to co-operate with each other and unified under the LCCs. The grassroots opposition, which lacked political experience, failed to launch an initiative that harmonised the efforts of political forces (Al Abed 2013). However, it did effectively develop a protest repertoire in towns and cities. As protesters overcame their fears and created ties with others, they also operated more comfortably on the Internet and the number of citizen journalists increased. At some point, they broke out of the state-imposed media blackout and bypassed the structural restrictions by satellite phones and Internet. Thus, as the weak repertoire of Syrians developed, the hybrid media activities of the protesters improved in parallel to these activities, de-bracketing the society-state nexus in Syria (Aslan 2015).

However, to peacefully sustain the action against an oppressive state required a well-developed protest experience which would bring with itself cooperation in all levels and resources. Such cooperation would have involved organisations that would connect the protesters with the outside world and create a tolerance culture. Chapter 4 showed that human right organisations and social movement organisations in Egypt were crucial to creating a tolerant atmosphere amongst the protesters. In the absence of these organisations, experienced activists in Syria tried to create this tolerant atmosphere through social media channels. However, as they did not get enough visibility, their voices could not be heard.

Notes

1. Interview with Rezan (pseudonym), Syrian journalist, Istanbul, 2013.
2. Interview with Abbas (pseudonym), Syrian activist, Istanbul, 2014.
3. Interview with Abbas (pseudonym), Syrian activist, Istanbul, 2014.
4. Interview with Osama, Syrian activist, Istanbul, 2013.
5. Interview with Osama, Syrian activist, Istanbul, 2013.
6. Interview with Mazen, Syrian activist Istanbul, 2014.
7. Interview with Mahmoud (pseudonym), the administrator of the Syrian Revolution page, London, 2013.
8. Interview with Mizyan Altawil, Syrian activist, Istanbul, 2013.
9. Skype Interview with Hediye Levent, journalist, 2012.
10. Interview with Mizyan Altawil, Syrian activist Istanbul, 2013.
11. Skype interview with Nadia (pseudonym), Syrian activist, 2014.
12. https://www.youtube.com/watch?v=kuit6PegrlU.
13. Interview with Mazen, Syrian activist, Istanbul, 2014.
14. Skype interview with Iyas Kaadouni, Syrian activist, 2013.
15. Interview with Dima (pseudonym), Syrian activist, Istanbul, 2013.
16. Interview with Osama (pseudonym), Syrian activist, Istanbul, 2013.
17. http://www.lccsyria.org/en/.
18. Interview with Mazen, Syrian activist, Istanbul, 2014.
19. Interview with Mizyan Altawil, Syrian activist, Istanbul, 2013.
20. Interview with Abed Alsarraj, Syrian activist and journalist, Istanbul, 2013.
21. Interview with Osama, Syrian activist, Istanbul, 2013.
22. Interview with Amer AsSayyed Omar, Syrian activist and journalist, Istanbul, 2013.
23. Interview with Hani (pseudonym), Syrian activist, Istanbul, 2013.
24. One example is the documentation website, ANA New Media Association: https://bit.ly/2WBbM5J.
25. http://souriahouria.com/category/espagnol/.
26. Interview with Rabee, Syrian activist, I exchanged emails with Rabee, 2014.
27. Interview with Mahmoud (pseudonym), the administrator of the Syrian revolution page, London, 2013.
28. https://www.facebook.com/pages/The-Syrian-Days-Of-Rage/1035790 39742154?fref=ts.
29. https://bit.ly/2FPp8FG.
30. https://bit.ly/2TQ8pWx.
31. https://bit.ly/2V9C92d.
32. Interview with Sama, I spoke on Facebook and exchanged emails with Sama, 2013.

33. Interview with SNC media employees, Istanbul, 2013.
34. Skype Interview with Caroline, Syrian activist, 2013.
35. Interview with Abbas (pseudonym), Istanbul, 2014.
36. https://bit.ly/2FNv95O.
37. Interview with Bassel (pseudonym), Syrian activist and journalist, Istanbul, 2013.
38. Interview with Abbas (pseudonym), Syrian activist, Istanbul, 2014.
39. Interview with Mizyan Altavil, Syrian activist, Istanbul, 2013.
40. Interview with Osama, Syrian activist, Istanbul, 2013.
41. Interview with Mahmoud (pseudonym), the administrator of the Syrian revolution page, London, 2013.
42. Interview with Mazen, Syrian activist, Istanbul, 2014.
43. Skype interview with Bayan, Syrian activist, 2014.
44. https://bit.ly/2FO1FDY.
45. https://www.youtube.com/watch?v=SFPVYTmHgjs.
46. https://www.youtube.com/watch?v=8EL4fgguGZo.
47. Interview with Bassel (pseudonym), Syrian activist and journalist, Istanbul, 2013.
48. http://www.youtube.com/watch?v=Uqj39ODLhzw.

References

Aday, S. (2012, October 10). *New media, old media, and the Syrian crisis* [Web log post]. Retrieved from http://takefiveblog.org/2012/10/10/new-media-old-media-and-the-syrian-crisis/.

Ajami, F. (2012). *The Syrian rebellion*. Stanford, CA: Hoover Institution Press.

Al Abed, T. (2013, March 18). Two years on Syrian revolution is now a civil war. *Al-Monitor.* Retrieved from http://www.al-monitor.com/pulse/politics/2013/03/syrian-revolution-drastic-changes.html#.

Al Assad, B. (2011, March 30). *President Bashar Al Assad's speech at the People's Assembly.* Retrieved from https://bit.ly/1xsISWr.

Al Barazi, Z. (2013). *The stateless Syrians.* Social Science Research Network. Retrieved from http://papers.ssrn.com/sol3/papers.cfm?abstract_id=2269700.

Almqvist, A. (2013). The Syrian uprising and the transnational public sphere. In C. Wieland, A. Almqvist, & H. Naddis (Eds.), *The Syrian uprising: Dynamics of an insurgency* (pp. 47–78). Fife, Scotland: St. Andrew Papers on Contemporary Syria.

Angelis, E. (2011). The state of disarray of a networked revolution. *Sociologica, 3,* 1–24. https://doi.org/10.2383/36423.

Aslan, B. (2015). The mobilisation process of Syria's activists: The symbiotic relationship between the use of information and communication technologies and the political culture. *International Journal of Communication, 9,* 2507–2525.

Athamneh, W., & Sayej, C. M. (2013). Engaging the authoritarian state: Voices of protest in Syria. *Journal of Arabic and Islamic Studies, 13*, 169–190.

Attacks on the press in 2011: Syria. (2011). Committee to Protect Journalists. Retrieved from https://cpj.org/2012/02/attacks-on-the-press-in-2011-syria.php.

Berenskoetter, F. (2012). Parameter of a national biography. *European Journal of International Relations, 20*(1), 262–288. https://doi.org/10.1177/1354066112445290.

Berger, L. M. (2008). The emotional and intellectual aspects of protest music. *Journal of Teaching in Social Work, 20*(1–2), 57–76.

Brewster, T. (2018, December 5). Syrian electronic army hackers are targeting android phones with fake WhatsApp attacks. *Forbes*. Retrieved from https://www.forbes.com/sites/thomasbrewster/2018/12/05/syrian-electronic-army-hackers-are-targeting-android-phones-with-fake-whatsapp-attacks/#783712706ce4.

Calhoun, C. (2001). Putting emotions in their places. In J. Goodwin, J. M. Jasper, & F. Poletta (Eds.), *Passionate politics* (pp. 45–57). Chicago, IL: University of Chicago Press.

CNN Wire Staff. (2011, February 12). The domino effect: Is there more to come in Middle East, Arab world? *CNN World*. Retrieved from https://cnn.it/31Xc3me.

Colla, E. (2012). *The people want*. Middle East Report 263: The Art and Culture of the Arab Revolts, 42. Retrieved from http://www.merip.org/mer/mer263/people-want.

Crisis Group. (2011). *Popular protest in North Africa and the Middle East (VI): The Syrian people's slow motion revolution* (Middle East & North Africa Report No. 108). Retrieved from http://www.crisisgroup.org/~/media/Files/Middle%20East%20North%20Africa/Iraq%20Syria%20Lebanon/Syria/108-%20Popular%20Protest%20in%20North%20Africa%20and%20the%20Middle%20East%20VI%20-%20The%20Syrian%20Peoples%20Slow-motion%20Revolution.pdf.

Della Ratta, D. (2018). Expanded places: Redefining media and violence in the networked age. *International Journal of Culture Studies, 21*(1), 90–104.

Della Ratta, D., & Valeriani, A. (2012). Remixing the spring! Connective leadership and read-write practices in the 2011 Arab uprisings. *Cyber Orient, 6*(1). Retrieved from http://www.cyberorient.net/article.do?articleId=7763.

Dijk, V. T. (1993). Principles of critical discourse analysis. *Discourse & Society, 4*(2), 249–283. https://doi.org/10.1177/0957926593004002006.

Eagleman, D. (2012, July 10). *Four ways the Internet could go down*. CNN. Retrieved from http://edition.cnn.com/2012/07/10/tech/web/internet-down-eagleman/.

Fowler, S. (2013, April 25). *Who is the Syrian electronic army?* BBC. Retrieved from http://www.bbc.co.uk/news/world-middle-east-22287326.

Free Syrian Translators. (2011, May 25). *The Syrian revolution: A one-year summary* [Web log post]. Retrieved from http://freesyriantranslators.net/2012/05/25/the-syrian-revolution-a-one-year-summary/.

Freedom House. (2012). *Syria: Freedom on the net 2012 report.* Retrieved from https://freedomhouse.org/report/freedom-net/2012/syria#.VdXs37JViko.

Ghrer, H. (2013, April). Social media and the Syrian revolution. *Westminster Papers in Communication and Culture, 9*(2), 115–122.

Goldstone, A. J. (2011). Understanding the revolutions of 2011. *Foreign Affairs.* Retrieved from https://www.foreignaffairs.com/articles/north-africa/2011-04-14/understanding-revolutions-2011.

Granovetter, M. (1973). The strength of weak ties. *American Journal of Sociology, 78*(6), 1360–1380.

Harkin, J. (2013, April). Is it possible to understand the Syrian revolution through the prism of social media? *Westminster Papers in Communication and Culture, 9*(2), 94–112.

Hassan, H. (2011, August 28). *Syria's opposition has failed to offer a viable alternative.* The National Opinion. Retrieved from http://www.thenational.ae/thenationalconversation/comment/syrias-opposition-has-failed-to-offer-a-viable-alternative.

Heacock, R. (2011, June 3). *Syria goes mostly offline as protests intensify* [Web log post]. Retrieved from https://opennet.net/blog/2011/06/syria-goes-mostly-offline-protests-intensify.

Holliday, J. (2011, December). *The struggle for Syria in 2011* (Middle East Security Report No. 2). Retrieved from http://www.understandingwar.org/sites/default/files/Struggle_For_Syria.pdf.

Index on Censorship. (2011, July 8). Silencing the voice of freedom in Syria. *Index the Voice of Free Expression.* Retrieved from https://bit.ly/2L42hrs.

Landis, J. (2011a, March 31). *Speech to the Syrian parliament by president Bashar Al-Assad: Wednesday, March 30, 2011* [Web log post]. Retrieved from http://www.joshualandis.com/blog/speech-to-the-syrian-parliament-by-president-bashar-al-assad-wednesday-march-30-2011/.

Landis, J. (2011b, April 16). *Assad speech II* [Web log post]. Retrieved from http://www.joshualandis.com/blog/assad-speech-ii-16-april-2011/.

Landis, J. (2011c, June 21). *President Assad: Speech #3 (Monday 20 June 2011) "We control events, rather than events control us"* [Web log post]. Retrieved from http://www.joshualandis.com/blog/president-assad-speech-3-monday-20-june-2011-we-control-events-rathern-than-events-control-us/.

Lesch, D. W. (2012). *Syria: The fall of the house of Assad.* Cornwall, UK: TJ International Ltd.

Local Coordination Committees. (2011). *Human rights violations committed by the Syrian regime: 15 March–15 October 2011.* LCC Media Office. Retrieved from http://www.lccsyria.org/wp-content/uploads/2011/11/SYR_HR_Violations_Mar15_Oct15_Final-3.pdf.

Local Coordination Committees of Syria. (2012, December 20). *Carnegie Middle East center*. Retrieved from https://bit.ly/2TVFLVC.

Lynch, M., Freelon, D., & Sean, A. (2014, January). *Syria's socially mediated civil wars* (United States Institute of Peace Report No. 91). Retrieved from http://www.usip.org/sites/default/files/PW91-Syrias%20Socially%20Mediated%20Civil%20War.pdf.

Macfarquhar, N. (2012, December 22). In ravaged Syria, Beach town may be loyalists' last resort. *NY Times*. Retrieved from https://www.nytimes.com/2012/12/23/world/middleeast/syrian-resort-town-is-stronghold-for-alawites.html.

Macleod, H. (2011, April 19). Inside Daraa. *Al Jazeera*. Retrieved from http://www.aljazeera.com/indepth/features/2011/04/201141918352728300.html.

Marsh, K. (2011a, March 23). Syria: Four killed in Daraa as protests spread across south. *The Guardian*. Retrieved from http://www.theguardian.com/world/2011/mar/22/syrian-protests-troops-kill-deraa.

Marsh, K. (2011b, April 1). Syrian security forces crack down on 'Friday of martyrs'. *The Guardian*. Retrieved from http://www.theguardian.com/world/2011/apr/01/syria-security-forces-crackdown.

Merrett, R. (2013, March 26). *Anonymous to protest against Facebook censorship*. Computer World. Retrieved from http://www.computerworld.com.au/article/457196/anonymous_protest_against_facebook_censorship/.

Mroue, B. (2011, March 23). 15 killed in clashes in southern city of Syria. *Washington Post*. Retrieved from http://www.washingtonpost.com/wp-dyn/content/article/2011/03/23/AR2011032300362.html.

Naggar, M. (2011, May 9). *Media at war in Syria*. Doha Centre for Media Freedom. Retrieved from http://www.dc4mf.org/en/content/media-war-syria.

Naidu, E. (2013, November). *Syria: Documentation and its role in memorialization*. Syria Justice and Accountability Centre. Retrieved from http://33bjj-c3l8q7dpgn6j1rg637f.wpengine.netdna-cdn.com/wp-content/uploads/SJAC-Documentation-Memorialization-Memo-2013_EN.pdf.

Najm, A. A. (2011, October 1). Syria's coordination committees: A brief history. *Al-Akhbar*. Retrieved from http://english.al-akhbar.com/node/764.

Noman, H. (2011, May 30). *The emergence of open and organized pro-government cyber attacks in the Middle East: The case of the Syrian electronic army*. Infowar Monitor. Retrieved from http://www.infowar-monitor.net/2011/05/7349/.

O'Bagy, E. (2012, April). *Syria's political opposition*. Middle East Security Report 4. Retrieved from http://www.understandingwar.org/sites/default/files/Syrias_Political_Opposition.pdf.

Officers fire on crowd as Syrian protest grow. (2011, March 20). *The New York Times*. Retrieved from http://www.nytimes.com/2011/03/21/world/middleeast/21syria.html.

Pies, J., & Madanat, P. (2011, June). *Media accountability practices online in Syria: An indicator for changing perceptions of journalism* (MediaCat No. 10). Finland: University of Tampere.

Reporters without Borders. (2012, March 12). *World day against censorship* (Internet Enemies Report 2012). Retrieved from http://en.rsf.org/IMG/pdf/rapport-internet2012_ang.pdf.

Rinke, E. M., & Roder, M. (2011). Media ecologies, communication culture and temporal spatial unfolding: Three components in a communication model of the Egyptian regime change. *International Journal of Communication, 5*(13), 1273–1285.

Shadid, A. (2011a, June 30). Coalition of factions from the streets fuels a new opposition in Syria. *The New York Times*. Retrieved from http://www.nytimes.com/2011/07/01/world/middleeast/01syria.html?_r=0.

Shadid, A. (2011b, July 21). Lyrical message for Syrian leader: 'Come on Bashar, leave'.*The New York Times*. Retrieved from http://www.nytimes.com/2011/07/22/world/middleeast/22poet.html?pagewanted=all&_r=0.

Shirky, C. (2009, December 11). The net advantage. *Prospect Magazine*. Retrieved from http://www.prospectmagazine.co.uk/magazine/the-net-advantage.

Silencing the voice of freedom in Syria. (2011, July 8). Index on Censorship. Retrieved from https://www.indexoncensorship.org/2011/07/silencing-the-voice-of-freedom-in-syria/.

Sinjab, L. (2011, March 4). *Syria: Why is there no Egypt-style revolution?* BBC. Retrieved from http://www.bbc.co.uk/news/world-middle-east-12639025.

Slackman, M. (2011, March 25). Syrian troops open fire on protesters in several cities. *The New York Times*. Retrieved from http://www.nytimes.com/2011/03/26/world/middleeast/26syria.html.

Smith, A. (2011, May 9). Syria blocks satellite phone communications. *Financial Times*. Retrieved from http://www.ft.com/cms/s/0/f255d800-7a69-11e0-af64-00144feabdc0.html#axzz3jdMjMoWr.

Stateless Kurds in Syria granted citizenship. (2011, April 8). CNN. Retrieved from http://edition.cnn.com/2011/WORLD/meast/04/07/syria.kurdish.citizenship/.

Syrian crisis: Troops move into towns in north. (2011, June 17). BBC. Retrieved from http://www.bbc.co.uk/news/world-middle-east-13802770.

Syrian electronic army: Disruptive attacks and hyped targets. (2011). *OpenNet Initiative*. Retrieved from https://bit.ly/2nHdDHR.

Syria protests: Homs city sit-in 'dispersed by gunfire'. (2011, April 19). BBC. Retrieved from http://www.bbc.co.uk/news/world-middle-east-13130401.

Syria protests: Making amends through the lens. (2011, July 5). *Integra*. Retrieved from http://www.integrallc.com/2011/07/05/making-amends-through-the-lens/.

Syria relaxes veil ban for teachers. (2011, April 6). *The Guardian*. Retrieved from http://www.theguardian.com/world/2011/apr/06/syria-relax-veil-ban-teacher.

Syria: Shootings, arrests follow Hama protest. (2011, July 6). Human Rights Watch. Retrieved from http://www.hrw.org/news/2011/07/06/syria-shootings-arrests-follow-hama-protest.

Tilly, C. (2006). *Regimes and repertoires*. Chicago, IL: University of Chicago Press.

Tobias, M. W. (2013, March 18). How and when to buy a satellite phone. *Forbes*. Retrieved from http://www.forbes.com/sites/marcwebertobias/2013/03/18/how-and-when-to-buy-a-satellite-phone/.

Tufekci, Z. (2014a, January 9). *Capabilities of movements and affordances of digital media: Paradoxes of empowerment* [Web log post]. Retrieved from http://dmlcentral.net/blog/zeynep-tufekci/capabilities-movements-and-affordances-digital-media-paradoxes-empowerment.

Tufekci, Z. (2014b, March 24). *Turkey's Twitter ban and social media demonization as a strategy*. Technosociology. Retrieved from http://technosociology.org/?p=1608.

Tufekci, Z. (2017). *Twitter and tear gas: The power and fragility of networked protest*. New Haven, CT: Yale University Press.

Van Dijk, T. A. (1993). Principles of critical discourse analysis. *Discourse and Society, 4*, 249–283.

Volkmer, I. (2014). *The global public sphere: Public communication in the age of reflective interdependence*. Cambridge, UK: Polity Press.

Williams, L. (2011, February 08). Syria to set Facebook status to unbanned in gesture to people. *The Guardian*. Retrieved from http://www.theguardian.com/world/2011/feb/08/syria-facebook-unbanned-people.

Zuckerman, E. (2011). Zeynep Tufekci on protest movements and capacity problems. *My heart's in Accra* [Web log post]. Retrieved from https://bit.ly/2ZuMOul.

CHAPTER 7

A Comparison of Leaders' Capabilities and Their Resources

In the 2011 Egyptian and Syrian protests, I identified three types of leadership. None of these leaders possessed formal legitimacy. They also did not appeal to the public with their charisma. Despite adopting an invisible leadership style, their vision and experiences influenced the organisation, ease and efficacy of the protests. The purpose of this chapter is to compare the strategies and capabilities of these leaders and to analyse the factors that affect the sustainability of the digitally supported movements they led. To enable a rigorous assessment of how leaders use the communication tactics in different political and social contexts and their capabilities during protests, I use the multi-layered framework outlined by Agarwal et al. (2014). This model tests three capability dimensions in today's movements:

(a) Resource mobilisation: The capacity to produce, allocate, and utilise material and symbolic goods that enable recurring patterns of collective action to occur. (b) Responsiveness to external conditions: The capacity to react to near term threats and opportunities resulting in asserting different repertoires of action, and assessing their results. (c) Long-term adaptation, change, or decline: The capacity to develop new patterns of association internally as the organisation grows, suffers resource loss, or receives less recognition by relevant publics. (Bennett et al. 2014: 234–235)

B. Aslan Ozgul, *Leading Protests in the Digital Age*,
Palgrave Studies in Young People and Politics,
https://doi.org/10.1007/978-3-030-25450-6_7

In this chapter, I explore if the leaders who led the Egyptian and Syrian protests demonstrated the three capabilities outlined above. While discussing the first and second capacities, resource mobilisation, I also refer to the narrative capacities of the movements, defined as the ability of a movement to frame its story on its own terms and spread it worldwide (Tufekci 2017).

Bennett et al. (2014) applied this framework to the 2011 Occupy protests in the USA and analysed whether they behaved as organisations and provide the necessary capabilities to sustain a movement. They found that although Occupy was formed by technology-equipped crowd networks and there was no main organising team behind the protests, it successfully operated within each theoretical dimension outlined above, as a networked organisation.

Like the Occupy protests in the USA, Bennett and Segerberg (2013) define the Arab uprisings as crowd-enabled movements. This chapter will also analyse and compare the organisational structures of the protests in Egypt and Syria and analyse the extent to which their leaders assisted in mobilising the resources, responding to external conditions and adapting the movements' structures to the changes. The findings presented in this chapter show that the Syrian and Egyptian protests essentially reflected different organisational structures. While there was a coherent and centralised organisation in the Egyptian protest, with the lead of a three types of leaders and a strong repertoire, the Syrian uprising reflected a decentralised structure with a weak repertoire. It was formed on the basis of the personalised action of experienced activists. I develop the theories of Bennett and Segerberg by explaining the reasons behind these organisational choices by the activists in Syria and Egypt, drawing on the approach of the leading social movement scholar Tilly (2006). Tilly, who analyses French protests in three periods of history, argues that in each instance, the French people made collective claims by drawing on the repertoire of contention available to them (Tilly 2006: 39). By tracing the protesters' activities in Syria and Egypt, I will show that the different organisational structure of the two uprisings appeared because protesters still drew on their respective repertoires of contention while taking part in the protests. The arrival of technology brought new opportunities to the protesters such as faster, more effective communication and new organisational methods, but could not completely change the parameters of the collective action. Adopting this technology in an effective way depended upon the past performances of protesters.

I also ask whether these structures affect the sustainability of the protests? Moreover, which of the other political and societal supports and constraints shaped the success and sustainability of the two movements? I define success criteria based on the definition of Gamson (cited in Jenkins 1983: 543), who measures success according to two dimensions: "(1) the provision of tangible benefits that meets goals established by the movement, and (2) the formal acceptance of the movement organisation by its main antagonists". The common goal of both movements was to mount a revolution, a rapid and systemic political change through a social movement (Cole 2014: 267). I will discuss what the sufficient factors to achieve this goal were. Most of the data for this chapter are drawn from both the history and empirical chapters of this book. I will also refer to the views of my interviewees, who were asked which factors differentiated the Syrian protests from the Egyptian ones.

Resource Mobilisation

Soft and Hybrid Leaders' Capabilities

My interviews with the Egyptian and Syrian activists showed that Facebook was the first technological tool that was used to mobilise the public in both countries. However, the experience of the soft leaders, their relationship with their followers, the framing they relied on and the operation of their Facebook pages showed huge diversity in accordance with the repertoire of contention of the respective country. In Egypt, the first medium used in the 25 January Revolution was launched by Wael Ghonim. This was the Kullena Khaled Said Facebook page (KKSFP), dedicated to Khaled Said, who was beaten to death by the Egyptian police in June 2010. Wael selected Abdelrahman Mansour as the co-administrator of the page, with whom he had already worked during the ElBaradei campaign. The two administrators had already acquired experience of leading a social media network via the Facebook page of this campaign in 2010. Wael and Abdelrahman knew that the identification of followers with the page was important (Ghonim 2012: 53). For this aim and security reasons, they stayed anonymous.

Abdelrahman[1] described the importance of anonymity for the administrators:

> If you stay anonymous, anyone can relate himself or herself to you despite your ideological differences.

The admins knew that online anonymity not only helps to avoid detection by the police, but also presents a discursive opportunity. They could use the anonymity to create different personas online than they exhibited offline. According to the administrators, the personality and ideology of the administrators are crucial factors that followers consider when associating him/herself with a Facebook page. Discursive opportunity scholars, such as McCammon et al. (2007: 731), also argue that the public looks favourably upon others who share similar ideologies. This creates the need for administrators to appeal to follower-centred ideologies in order to persuade them to side with the movement's members (McCammon et al. 2007: 731). Being online helped Wael and Abdelrahman to not reveal their identities and ideological backgrounds. They could echo and amplify followers' specific ideologies. For example, the administrators portrayed themselves as ordinary Egyptians not working for any organisation, political party or movement of any kind. They dedicated the page to Khaled Said, who resembled an ordinary follower of the page, a middle-class citizen. They spoke in the name of Khaled on the page under the pseudonym "El Shaheed" (the Martyr). They also rejected the political activists' language and adopted the colloquial Egyptian dialect and non-ideological discourse (Ghonim 2012: 61). This lack of specificity of the administrators, their choice of language and their non-political stand while framing the problems with the regime helped the identification of the page members with Khaled Said (Herrera 2014: 52).

As in Egypt, in Syria three friends took the initiative and launched the Syrian Revolution Facebook page (SRFP) from outside Syria on 18 January 2011. I asked one of these administrators, Mahmoud[2] (pseudonym), whether the KKSFP influenced them when they launched their page. He responded that:

> I knew most of the Egyptian activists and I believe that their experiences were very different from ours. We just liked their idea that we could use Facebook to get connected and communicate with youth and be near to their eyes. This is the platform that youth likes.

Mahmoud believed that their experiences were different from the experiences of the Egyptian activists as the regimes in two countries were different and according to him, "launching a Facebook page against the Syrian regime was a life and death decision as all the systems were monitored by the regime" (interview with Mahmoud, 2013). The words of

Mahmoud revealed his fear of the Syrian state, whose presence could be felt in every aspect of life and social interactions (Al Hendi 2012). Despite having launched the page, Mahmoud was feeling a low level of self-efficacy due to his discomforting experiences of the Syrian protesters in the past (Human Rights Watch 2009). Like Mahmoud, fear was the emotion that was frequently noted in an explicit way by the Syrian interviewees for this study. 15 of 22 Syrian activists interviewed used apprehensive words such as "fear", "afraid" or "scared" during the interviews.

Unlike the SRFP's administrators, Abdelrahman,[3] the second administrator of the KKSFP, was an experienced activist who has already been subject to torture. When I asked him whether he was afraid of being arrested, he responded:

> I should be awake and break this fear inside of me first to break fear inside others as well. I was alleviating my fear through writing on social media. Also, I wasn't afraid of torture. I have been in prison three times. Yes, my body was in pain but when you demolish the power of the state in your mind, they can no longer represent something strong to you.

The past activism of the administrator taught him how to break the fear in him but also in others through social media. Unlike the SRFP which was launched two weeks before the call for protests, the KKSFP was launched six months prior to the 25 January Revolution. The administrators had more time to mitigate fear. They used silent standings as a mechanism of communal gatherings where members met with each other during these six months. The purpose behind the silent standings was to activate two mechanisms that mitigated the fear. The first was to create social networks that underpinned the 25 January Revolution. The second was to boost the identification of the activists with the page that would activate members to take on a difficult task and sustain them during periods of scant success during the 25 January Revolution (Ghonim 2012: 73).

On the other hand, unlike the KKSFP administrators, this was the first time that Mahmoud and his friends managed a social media page against the Syrian state. Lacking administrative experience, they had difficulty managing fear, which has been stressed as an important success factor by social movement theorists Goodwin and Pfaff (2001: 284). To understand the role of emotions in the movements, Goodwin and Pfaff analysed and compared the US civil rights movements of 1950 and

1960 and the East German civil rights movement of the late 1980s. They found that there were six encouragement mechanisms that were operative in both movements to mitigate fears. These were: "(1) the intimate social networks that underpinned these movements; (2) the dynamic of mass meetings or other communal gatherings of movement participants; (3) the strong identification of activists with the movements; (4) shaming and degradation ceremonies; (5) formal training in the techniques of civil disobedience; and (6) mass media coverage of movement activities and protest events" (Goodwin and Pfaff 2001). In the SRFP case, mass media coverage increased the number of people who liked the page, but it could not mitigate fears in the public by itself. The symbol used on the page (a clenched fist and red colour), the religious language adopted by the administrators, the date chosen for protests reminding the Muslim Brotherhood protests in Hama and the selected spokesperson of the SRFP, who was a member of Muslim Brotherhood, created doubts in the mind of the public and prevented the identification of secular followers with the page (Aslan 2015). The frame that protesters relied on is critical in providing rationales for potential supporters to side with them (McCammon et al. 2007: 728). In the case of Syria, as the online framing capabilities of protesters had not yet developed, they first selected rhetorical strategies that do not align well with the popular beliefs of their followers. Hence, it is seen that the effectiveness of the Internet as a tool depended on the capabilities of these soft leaders.

When the organisational structure of the KKSFP and the SRFP was compared, the SRFP's administrators were found to be composed of 40 people from across the world. Mahmoud[4] mentioned that they chose these members from amongst the followers who actively participated in the discussions on the page. Apart from the founding administrators, the others did not know each other's identities due to security reasons, he continued:

> The administrators should be together in order to decide what is good and what is not. In our case, we did not have this opportunity. Each week, we had to talk online about each political situation, inside or outside of the country, to decide what approach we should take because it was impossible to meet. I was meeting daily with the other person who was managing the web page with me but, for the others, we all had to be in the same place. So, the only way to do this was to write a post in our closed Facebook group and discuss there.

Mahmoud's words show that although the founding administrators decided who could join the administrative board, they preferred a democratically organised approach and included other administrators to create strategy and frame issues. Yet, being in different countries restricted the framing capacity of the administrators to react to urgent issues. As these 40 administrators were in different time zones, they often had to wait before getting the perspectives of other administrators on posts and sharing them with their followers. This slowed down their activities on the page.

Wael and Abdelrahman also adopted a democratic approach and created a framing strategy together on the KKSFP. Abdelrahman[5] said:

> The interaction of me with Wael was like; I was the one who was pushing and he was the one who was pulling. This balance created a good vibe on the page. It was not too radical but also not too soft.

Yet, on 27 January 2011, when Wael was arrested and Abdelrahman went to the army, Sami (pseudonym), became the sole administrator of the KKSFP during the 10 days of the 25 January Revolution and continued to help Wael for a short period afterwards. According to Sami,[6] Wael was influential in determining the discourse that was used on the page during that period:

> Each of us had our own preferences, but in terms of discourse, it was Wael who was the one in charge and had the final say.

The statement of Sami was important in indicating that the framing on social media was not always structured in an egalitarian and horizontal manner, as we would expect from the organisation of crowd-enabled connective actions. Conforming to the "soft leader" concept of Gerbaudo, Wael controlled the flow of information and influenced the followers and the third administrator with his discourse.

Another difference between two Facebook pages was that, apart from Wael, the administrators of KKSFP, Abdelrahman and Sami were in Cairo in 2011 and could join the meetings on the ground. They participated in person in the silent standings organised by their page and be in touch with the group that would later be called the Revolutionary Youth Coalition through email exchange.[7] Drawing on their past protest experiences, administrators acted as hybrid leaders that mobilised the movement on the ground but also on the Internet. They had an influential and mobilising effect in both spheres.

This is something that we did not observe in Syria where protesters adopted precautions while using of ICTs. The experienced activists that I interviewed for this study mentioned that they called each other by phone and used secret words to call for protests. They also used social media channels for interpersonal communication, not just because of the communication culture in the country but also due to security concerns. Instead of launching an open Facebook page where they could reach millions, they opened secret Facebook or Skype groups and talked with friends or friends of friends through these pages. Hence, they used social media as electronic word-of-mouth rather than as a space for open public discussions.

The lack of open public discussions prevented open peer production, in which large crowds created and shared content through social media, phones, SMS and email (Bennett et al. 2014: 235). It also prevented the emergence of hybrid leaders who would influence the public and act as mobilisers and coordinators in both the online and offline spheres. The slow Internet speed was also an obstacle for the peer production mechanism. The Syrian diaspora took an active role in sorting, distributing and drawing timely attention to the protest news and videos on YouTube and Twitter. However, this peer production mechanism of the diaspora only informed the outside world and could not mobilise the protesters inside Syria. Hence, as Tilly (2006: 40) cites, past familiarities increase the likelihood of subsequent performances and enlarge the social networks of the protesters. As the Syrians lacked past protest experiences, they could not benefit from large social networks in which they would trust, produce and mobilise the protest messages from person to person on the Internet. External factors to the protesters such as the Internet structure were another constraint that limited their capabilities.

On the other hand, in line with past protest experiences, the capacity of the Egyptian activists to produce, allocate and utilise the Internet was more developed than that of their Syrian counterparts. They used this tool to frame the problems with the current regime, create an identity amongst Internet users, encourage them to take to the streets and gain protest experience before the 2011 protests broke out.

Experienced Activists' Capabilities

As Sidney Tarrow (2011) claims, movement organisations in a country progress in interaction with cultural artefacts, power holders and other movements (Tarrow 2011: 127). Like organisation, leadership skills have to be learned. The education and trial and error experiences of activists are crucial to teaching leadership skills as a movement unfolds (Oberschall 1973: 58). As explained in Chapter 2, Egypt saw several incidents of collective action on a daily basis since 1952 (Abdelrahman 2014). Unlike Syria, it also enjoyed multiparty presidential elections. Although the elections in Egypt were not as democratic as they may have seemed, they paved the way to the rise of different political figures, such as ElBaradei, who would bring hope for change to Egyptians in 2010. The activists who worked for the campaign of ElBaradei acquired organisational capabilities in both the online and offline spheres. The repertoire of contention in Egypt, with movements and political campaigns, gave way to the formation of the Revolutionary Youth Coalition in early January 2011, which went on to organise the 25 January protests (Levinson and Coker 2011). The Revolutionary Youth Coalition's members obtained leadership experience as the movement unfolded in Egypt from the 2000s. The group was composed of the representatives of political movements, organisations and party members that were once working separately (Shehata 2011). Although they have never claimed to be the leaders of the movement, they acted as the leaders and planned the times and places of the marches that took place during the 2011 Egyptian protests. Mona,[8] one of the representatives of the Revolutionary Youth Coalition, describes the mobilisation process as dividing into groups of five and six people and each group informing a specific neighbourhood. They mostly used face-to-face communication and also distributed brochures and leaflets to citizens. She emphasised the importance of offline mobilisation:

> Online was useful but it was not enough to mobilise people. What was more important was going to the neighbourhoods that had problems with the police. For instance, I went to the Omrenea. This was an area where Christians had big problems and many were previously killed here because of the bombing of the church. When I went there and told them that they should go out on the 25th, they said they would go. They mobilised the whole neighbourhood to join in the protests.

Similarly to the Youth for Change (Al Shabab min Ahl Al Tagheer), whose members targeted the working-class Cairo community and visited public squares or parks in groups of four or five people in 2005, the Revolutionary Youth Coalition engaged with the public in different neighbourhoods (Azimi 2005). In particular, they succeeded in activating citizens who did not have access to the Internet. These experienced activists on the ground convinced the poor segments to join the protests by addressing their problems and asked to protest with them for bread, jobs and social security.

The organisation of the Youth Coalition also went public through social media channels. The organising group cooperated with the KKSFP administrators and benefitted from a large number of followers of that page. Mona[9] said:

> Khaled Said was the main page. The other pages were getting the links from there and publishing them as well. The reason for this was, the KKSFP was the only anonymous page back at that time. The authorities knew the administrators of other pages. The KKSFP also had a good reach with thousands of followers.

Mona also mentioned that, along with the KKSFP, they used the ElBaradei Facebook page and Twitter to distribute specific posters for the Egyptian revolution to appeal to middle-class Internet users. They also performed unfamiliar practices such as distributing incorrect information about the protest sites on the Internet in order to baffle police officers by misdirecting them to non-protest sites during the 2011 Egyptian protests. Hence, it was difficult to separate the action on the ground from the action on the Internet, the organising group successfully combined both online and offline mobilisation techniques to reach different segments of the society. According to Tilly (2006: 40), if a movement strongly adopts familiar performances and also benefits from unusual practices, the movement's members come to prefer the flexible repertoire, which he calls "strong". In the case of the Revolutionary Youth Coalition, along with familiar performances, the organising group adopted new mobilisation techniques, which can be called a strong repertoire of contention on the ground.

On the contrary, the lack of past protest experiences led Syrian activists to adopt a weak repertoire. In weak repertoires, the past familiarity increased the likelihood of subsequent performance in a more or less linear manner. It reflects the effects of learning but not of strong preferences (Tilly 2006: 40). What sparked the protests in Syria was the grievance and repression in Daraa. The citizens of Daraa first witnessed the arrest and torture of 15 children who drew anti-regime graffiti on the wall, and then, skirmishes erupted when a small number of people marched demanding the release of the children (Leenders 2012). According to Leenders (2012), who explored the dynamics and underlying conditions of Daraa, the close social networks of Daraa citizens which appeared due to cross-border trade between Daraa and Jordan, as well as the crime network in the city, facilitated the emergence of a collective identity amongst Daraa's citizens. On 25 March, the protests spread across Syria and activists who had protest experience took the lead in cities. As one of the leading activists from Darayya, Osama[10] reported that they divided into groups of five or six to mobilise people, and after creating large numbers, they gathered in the mosques. Hence, similarly to the Revolutionary Youth Coalition of Egypt, they mobilised the public with snowballing mobilisation techniques.

However, as most of the activists did not have any previous experiences due to the brutal repression of the regime in the country, they feared of being caught. They prudently acted while contacting with others. This led the Syrian activists to use the offline repertoire in a restricted way too. Abed,[11] for instance, summarised the situation, saying:

> You cannot trust anyone, even though you present together in the protests.

This fear was related to the lack of protest networks and protest experience in the political sphere. As Abu Najm claimed, Syrians were "reduced to secrecy and denied the right to organise themselves for almost five decades" (Najm 2011). Consequently, they had to build their repertoire of contention from scratch. Unlike the Egyptians who built their social networks over the past years of protests and campaigns, the Syrian activists created their protest networks only during 2011. Hence, when they first went out to protest, they did not know whom to trust. Unlike the Egyptian activists who mobilised the public in different neighbourhoods, the Syrian activists only spoke to and mobilised people that they already knew, due to the fear they sensed. Each person in the group brought friends or friends of friends to the protests.

They could therefore not adopt flexible mobilisation techniques that would give them the advantage of mobilising the public more quickly. The lack of past protest networks further slowed the mobilisation process. Thus, the Syrian case disclosed the fact that the low number of experienced activists, as well as protest networks in Syria and the broader social and political context in the country, restricted the resource mobilisation capabilities of the activists.

Responsiveness to External Conditions

State repression is one of the most important external conditions that the movement needs to respond. Before the start of 2011 Egyptian protests, the central organising body, formed by the members of the Revolutionary Youth Coalition, created a centralised and well-planned organisation to resist the state repression. For this, they used two important elements in Egypt. The first was the presence of a hybrid media system, which was shaped by the interaction of online and mass media channels such as Al Jazeera and Al Masry Al Youm (Chadwick 2013). Some individual participants learned about the protests thanks to this hybrid media channels and went public with the organising group (interview with Mahmoud, 2014). The second mechanism was the co-ordination of the Revolutionary Youth Coalition (the organising group on the ground) with the KKSFP. The KKSFP became a stitching mechanism for the co-ordination process, which linked the Internet users with activists on the ground. The protest time and spaces were shared by more than 50,000 people through this page (Ghonim 2012: 166). On 25 January, when the protest groups started to march in small streets, the members of the organising group co-ordinated with each other through mobile phones (interview with Mona, 2014). This reminded the "Global Justice Movement" in Egypt whose members also co-ordinated with mobile communication and converged on specific locations. Twitter was also an important co-ordination tool. Sherif Azer,[12] an Egyptian human rights activist, said they learned how to use Twitter for the protests during the workers' movements on 6 April:

> The first time I started to use Twitter, I realised its importance. It was during the protests on 6 April, 2008. Back then, we needed to send an SMS to an international number in order to send a tweet. We were trying to integrate with people from Mahalla El-Kubra. There were activists that were trying to go in and the police stopped them. We were updating each other on which way was safe to go.

In 2011, with the improved instantaneity of Twitter, the activists covered events as they happened and connected with each other via this new channel (Papacharissi and Oliveira 2012: 9). On the basis of the information coming from ICTs, independent groups operated in line with each other and marched towards Tahrir Square. The organising group was also ready for the possible blackout of the Internet. During our interview in 2014, Mona,[13] a member of the Revolutionary Youth Coalition, mentioned that they were expecting that the regime would slow down the Internet connection, so they all exchanged their landline numbers. They also prepared and printed Google map data prior to 27 January in anticipation of a possible Internet blackout (interview with Esraa Abdel Fattah, 2014). On 28 January when the Internet was shut down, they thus easily adapted their activities to the changing structure of the Internet.

The experienced activists were key to the smooth continuation of the movement on the ground. Protest groups gathered in mosques and moved in co-ordination via the Google Maps, which was prepared in advance and distributed via ICTs.[14] The long-established political groups, such as the Muslim Brotherhood, also worked in co-ordination with the Revolutionary Youth Coalition. Mona,[15] for instance, indicated that:

> Even though some activist groups were not in the Revolutionary Youth Coalition, we all knew each other and they also worked with us. We started knowing each other more after January 25. We had a meeting with all groups at 6 pm every day. We were deciding what we would do the next day in Tahrir, so it was actually like managing Tahrir.

The cooperation of the Revolutionary Youth Coalition with the KKSFP and other political groups in 2011 helped generate the appearance of a centralised organisation—exactly what had been absent in Syria. The interviews with the activists who mobilised the protests in Mansoura and Alexandria showed that experienced activists in peripheries were in contact with activists in Cairo and acted in line with their strategies. John,[16] an activist in Alexandria, for instance, mentioned that they also started the protests in the small streets of Alexandria and worked in co-ordination with the activists in Cairo. Thus, despite the changing structures on the Internet, the Revolutionary Youth Coalition implemented coherent strategies and controlled the protests with the help of the KKSFP, other political groups and newspapers.

Unlike the Egyptians, the Syrian activists followed a decentralised structure. Local protesters planned the online and offline tactics to bypass the security and mobilise the protests. The experienced activists who unified under local committees used to gather in coffee shops during the day and talked in closed Facebook and Skype groups at night. They planned chants, slogans and banners in each city.[17] Yet, the protests never became as co-ordinated as the centralised Egyptian movement. The researcher Enrico Angelis[18] described the protests as follows:

> These movements inside Syria were much separated. Different actors, different slogans… It was very fragmented. If you go to Edlip, you will see that it was different than Haleppo or Daraa.

Moreover, in the absence of past protest experiences, the Syrians were not ready to adapt their actions to the structural changes brought on by the Syrian state. Like the Egyptians, the Syrians used the mosques as gathering points and simultaneously started the protests after Friday prayers. However, they were not ready against the state repression when they went out of the mosques. During the first protests across the country on 25 March, the protest groups that started marching from the mosques did not know what to do or where to go after leaving them.[19]

The lack of open hybrid media and the discourse of the Syrian state was also a constraint that the protesters had to bypass. As discussed in Chapter 6, unlike the Egyptian state, the Syrian state was consistent in its message throughout the different stages of the protests. Both the president and national media consistently used sectarian discourse, claiming that the protesters comprised extreme armed groups that threatened the unity of the country (Athamneh and Sayej 2013). The regime used this strategy on the online sphere too. Instead of shutting down the Internet completely during the protests, as the Egyptian state did, the Syrian state would only cut access to the Internet in specific cities for limited periods and mobilised its cyber warriors to counter-attack. The cyber warriors who spread the alternative narrative of the state would deny the protesters control of the online narrative (Harding and Arthur 2013). As Tufekci (2017: 231) claims "Without traditional trusted institutional gatekeepers, it is quite difficult for an ordinary person to know what is true and what is a hoax, or who is reliable and who is untrustworthy on the Internet". Using this advantage in the online sphere, the state attempted to lower trust towards the protesters by enforcing a sectarian narrative at all media channels. The repetition and persistence of the messages had an impact on minority groups, such as Alawite, Shia Muslims and Christians and on urban elites who feared instability. Some withdrew their support of the protests over time.[20] In the absence of hybrid media opportunities, the protesters had difficulty responding to these threats. Media restrictions blocked the information flow to the outside world and limited the dissemination of news across Syria (Naggar 2011). In Chapter 6, it was also noted that during the army's occupation in April and May 2011, Syrians could not get information from other cities under siege. In Egypt, we had seen that political and non-governmental organisations played a significant role in spreading information to the outside world during the Internet shutdowns. The Syrians, on the other hand, could not respond rapidly to these challenges posed by the state due to the absence of established political and non-governmental organisations. Only when they acquired experience in protest fields and developed social ties did they learn how to bypass state restrictions via satellite telephones and the Internet. These mediums were provided by Syrian media agencies outside the country.[21]

Long-Term Adaptation, Change or Decline

As Tilly says, "the outcome of collective action depends very strongly on the course of interaction" (1985: 718). This analysis has shown that this rule does not change in digitally supported movements. An important factor that provided sustainability to the 25 January protests in Egypt was the synergy between the leaders in the streets and soft leaders on the Internet. This interaction and eventual coalition-building and unification of different political groups started with the second Palestinian uprising in 2000 and developed with Kefaya (Abdelrahman 2014: 34). Kefaya was able to attract disparate parts of Egypt's opposition by spreading to 24 of the 26 provinces ("The Kefaya Movement" 2008: 18). It also created a group called Youth for Change that particularly targeted the young generation and linked activists on the streets with the middle-class segment on the Internet (Sha'ban 2006: 135). In 2008, Egyptians experienced a new coalition attempt when the workers of El Mahalla El-Kubra called for protests and activists from outside the city offered support. The April 6 movement Facebook page administrator, Esraa Abdel Fattah, was one of these activists, who brought support to the workers. She mobilised the protests on the Internet, but also organised the protests in the Cairo streets. Although the protests in 2008 were restricted in El Mahalla El Kubra and could not spread to Cairo, the organising group learned how to approach workers and motivate them to take to the streets (interview with Esraa Abdel Fattah, 2014). Activist cooperation and unification was also seen online. Bloggers across the political and religious spectra expressed solidarity with each other (Radsch 2008: 7). All of these cooperation and unification experiences of the leaders helped to build a centralised movement during the 2011 protests.

When the organising group Revolutionary Youth Council started to organise the 25 January protests, they first received support from KKSFP administrators, who would spread protest information amongst Internet users and regularly update them about the street protests during the revolution (interview with Sami, 2014). Certainly, having different types of leaders complicated the decision-making process and led to the appearance of conflicting ideas. This conflict became particularly apparent in the memories of Ghonim (2012), who mentions that the starting point of the 25 January protests became a serious matter for the leaders. As a soft leader, Ghonim did not at first agree on the starting point of the protests with Ahmed Maher, the co-founder of the April 6th movement and an active figure on the ground. Ghonim mentions, drawing on his memory, that:

> Ahmed Maher suggested that demonstrations begin from the Ministry of interior or in Tahrir Square. I opposed this idea strongly. To me these locations typified traditional demonstrations venues. I also spoke with Mahmoud Samy about locations. He was very excited about Arab League street…I know I was not a field expert so I deferred to Mahmoud Samy and Ahmed Maher among other activists, and asked them to coordinate with each other.

In his recollection, Ghonim accepts that as a soft leader living abroad, he lacked expertise in the field. This inexperience led him to agree with the important positions of the experienced activists on the ground and to their deciding on a key issue. Learning from their past protest experiences, the leaders knew well that cooperation is key to the sustainability of protests. In order to achieve their collective aim, they mustered their forces over the 18 days.

On the other hand, the conflict of interests prevented certain groups from collaborating with the protesters on the ground. Afraid of losing their power, Muslim Brotherhood leaders had been reluctant to support the protests in the first days of the protests. Yet, although the leaders refused to support the 25 January Revolution at the start, the Youth Council used the social and financial networks of this long-established group. For example, as a member of the organising group in Alexandria, John Albert[22] claimed that:

> The ElBaradei campaign had only 10 or 15 people in Alexandria. There were also people from the April 6 movement but, actually, we used to work with Ikhwan, which was really effective. They possessed a huge number of people. They were asking us how many people were going to the protests. We, for example, were saying 15 and they were sending 15 people as well.

Thus, the Muslim Brotherhood's social and financial networks helped the activists to enlarge the protests' scope. Moreover, human rights organisations such as the Hisham Mubarak Law Centre helped to sustain the protests by updating the outside world about the ongoing protests in order to erode support for the Mubarak regime (Rogin 2011). The cooperation of the youth that appeared over the years of online and offline activism also helped to sustain the crowds in Tahrir (Gunning and Baron 2013: 260). However, Chapter 4 of this book showed that, when Mubarak gave his final speech and promised that he would not run in the next elections, the support for protests started to erode. While some Egyptian revolutionaries tried to keep the dissidents in the square, others began to evacuate the square feeling that their aim had been fulfilled. At this point, the hybrid leaders were effective in finding new patterns of action to gather and renew the support. They used sarcasm and tried to delegitimise the actions of the state online. They also constantly updated the public by sending photos and information from the square. What brought support to the protests was also the excessive use of power by the state. As explained in Chapter 4, the "camel battle" brought energy to the protests. Another significant factor that helped the protests grow was the participation of worker syndicates. With the contribution of workers, life in small but also large cities essentially stopped in Egypt (Gunning and Baron 2013: 181). A final factor that fostered the sustainability of the Egyptian protests was division amongst elites. As the dominant interests of the army clashed with those of Mubarak, the army did not interfere with the protests until the last minute (Salem 2013).

When the protesters succeeded in stopping economic life in Egypt, the centralised structure of the country meant the army could easily replace Mubarak and retain his regime (Stacher 2012). Hence, the Egyptian case showed that there was a combination of factors that helped ensure the sustainability of the protests, leading to the overthrow of Mubarak and forcing the government to formally accept the protests.

On the other hand, in Syria, as the protest experience of the activists improved, cooperative initiatives against external threats to the protests emerged, such as the Local Coordination Committees. These committees helped to form digital networks between activists in different cities and prepared them for possible threats that they might encounter (Local Coordination Committees of Syria 2012). However, unlike Egypt, the peaceful protests in Syria turned into a military conflict six months after the uprising began. According to the Syrians interviewed for this study, the brutal repression of the regime, lack of support from the international community and lack of past political experience were the factors that changed the process of protests and pushed some activists to take up arms. For instance, due to the army's brutal response, activists could not occupy Clock Square in Homs and draw the attention of the international media to that area, as the Egyptians did in Tahrir Square (BBC News 2011). The army's brutal response was related to the regime's structure. As outlined in Chapter 2, President Bashar al-Assad initially had to share power with different power centres distributed amongst the party, the cabinet and the army and security forces. However, the key leadership positions of the army were filled with the Alawi officer corps who were from the same sect with President al-Assad (Zisser 2001: 19–20). Fearful of being pushed out of power, the Syrian army immediately opposed the protests and supported the regime (Landis 2011). Although certain army officers and soldiers later left the army and tried to impede its violent practices against opposition groups, in the first months there was no division amongst the elites (Haddad 2012).

So far, I refer in particular to the state-related factors that restricted the capabilities of the Syrian activists to sustain the peaceful protests; however, another important factor that was mostly articulated by the experienced activists amongst my interviewees was the inexperience of Syrians in forming and organising opposition groups. For instance, Mahmoud,[23] the administrator of the SRFP, emphasised the fact that Syrians had had no chance to be exposed to political life in the last 50 years and that this had diminished their capacity to cooperate with each other. The lack of good protest organisation was also sensed between the big cities and peripheries. First, this failure was due to the significant military presence in Damascus and Aleppo (Lesch 2012: 96). Second, these cities' middle and upper classes traditionally refrained from politics and did not want to lose their economic, social and political benefits (Lesch 2012: 96). For instance, Caroline,[24] the founder of Radio Souriali, an opposition online radio station, was a good example of these citizens living in the big cities. As a Christian living in Damascus, this young businesswoman said that:

> I had a perfect job and great life. I did not know what life outside Damascus was looking like. In January [2011], one of my good friends asked me if an uprising would happen in Syria. I said no because I was fine in Syria, I did not know what was actually going on in other parts of Syria.

An active and diverse blogosphere as the Egyptian one might have played a significant role in maintaining this segment's engagement with politics by raising the political consciousness of the society.

An active blogosphere might also appease the fear of chaos of the urban elites, who worried about the centrality of Daraa in the protests to the protesters. Daraa was known for its tribal loyalties, poverty and Islamic conservatism. This made the protests there lose appeal to the urban elites. Even though these elites shared anger and hope for liberation with Daraa's people, they feared the poor and the threat of disorder (Landis 2011). Hediye Levent,[25] a journalist working in Damascus, said:

> The elites have been present in the first protests in Damascus but seeing protesters coming out of mosques by chanting Islamic slogans withdrew them from the streets.

As explained in Chapter 2, thanks to their borders with Lebanon and Iraq, Syrians could see how sectarian tension could violently destroy the fabric of society (Lesch 2012: 50). Due to their inexperience, activists frame their grievances without regard to dynamics in the broader cultural context. The Islamic slogans adopted by the protesters were neither sensible nor legitimate to the political beliefs of the urban elites in the big cities. Instead of facilitating the reception of protesters' demands, they fortified the long-lived ideological discourse of the state, which often drew attention to the enemies that always stood at the gates and threatened the safety of the homeland. The lack of support from the big cities and its urban elites diminished the power of the protests.

Finally, the arrest of experienced activists in rural areas further challenged the endurance of the peaceful protests. For instance, Osama, an experienced activist from Darayya, was arrested at the beginning of the protests. He told me that, when experienced activists like himself came out of prison, the protests had already become militarised. Osama[26] continued:

> The regime was clever. They arrested everyone, especially the elites. Only uneducated people were left in the streets and they did not have the same values of the elites. Syrian society is very conservative. When Assad forces went to a place, they provoked people. People went out of control. The elite could have controlled these people and they could have made them aware of the danger of what they were doing. But, the elites were in prison.

Osama explains the appearance of violent protests in Syria with the conservatism and lack of leaders. According to Osama, the educated and politically active public who were arrested in the first months of the uprising would act differently than the protesters who continued to protest on the streets. Agreeing with Osama, another interviewee for this book, Bsher, also mentioned that with the arrest of elites, the protests were led by villagers and imams from rural areas. These new leaders did not know how to counter-frame the state's claims.

As Chapter 6 explains, the state used the "victim discourse" from the start of the protests, claiming that these were armed protests and the security forces the real victims. It blamed the protesters for the chaos in the country. Similar to the Egyptians, the Syrian activists could use sarcasm and tried to delegitimise the actions of the state by adopting a unifying and secular counter-framing in persuading the public (McCammon et al. 2007: 735). Instead, the religious discourse of some opposition protest groups and the transformation of the peaceful protests into armed conflict fortified the victim framing of the regime. As Hanisch (2001: 79) points out, leadership is a matter of having enough vision to show the way and enough verbal agility. As the activists did not have enough experience, they could not heighten the movement's chance to convince the elites in urban centres. This once more proved the importance of leaders and their capabilities for the sustainability of the digitally supported movement. The findings show that, leaders are still crucial in today's movements to increase the political awareness amongst the middle-class segment, connect the urban cities with rural areas, unify protesters around secular, non-ideological aims, create co-ordination and coherence amongst protest groups and prevent the extremism to emerge in line with state repression.

CONCLUSION

In her inspiring book, *Twitter and Tear Gas*, Zeynep Tufekci (2017) mentions that network internalities, which are the collective capabilities attained during the process of forming durable networks, such as building trust and delegation amongst a semi-durable network of people, are key to the sustainability of today's movements. The 2011 Syrian protests showed that the network internalities were not only necessary to sustain the movement, but were also crucial in igniting a movement in a non-democratic state. My analysis shows that in both the Egyptian and Syrian protests, "trust relations" were vital for connecting and activating networks. In Syria, the long-standing coercion structures such as violence towards the protesters and the regime's semiotic practices alienated the public from political life. In the absence of pre-existing social networks, loose ties formed on a Facebook page could not replace "trust-based", direct contact. As Tilly (2006: 55) argues, the relative efficacy of tools depends on the match of tools, tasks and users. Lacking protest experience, the soft leaders of Syria could not mitigate the fear amongst Internet users and motivate them to join in the protests. Instead, the experienced activists who had already established dense social networks and trust relations on the ground played the key role of activating the public. As they were afraid to be caught by the authorities, they mostly used ICTs for interpersonal communication.

On the other hand, the Egyptians have already built trust networks and delegated specific roles to particular activists before the protests started on 25 January. The experience of them in online and offline activism also made them efficient in incorporating ICTs into their repertoire during the 2011 protests. Drawing on their past protest experiences, they knew that creating a collective identity amongst the Egyptians would only be possible by staying anonymous. Considering the threats posed by the state, this was also the most secure method. Using ICTs effectively, soft and hybrid leaders benefitted from the ICTs' advantages and created new identities on the Internet. As was explained in Chapter 3, those like Asmaa Mahfouz who attempted to have an impact by showing their face on social media were the exception (Wall and Zahed 2011: 1338). When the structure of two social media pages was analysed, it was seen that ICTs encourage horizontal and participatory decision-making structures. However, having a large number of administrators in different time zones created difficulties in responding to the challenges posed by the state, resolving the disagreements and restricting the organising capacity of the soft leaders.

The resource mobilisation capabilities and responsiveness of the protesters to external conditions were also dependent upon the past protest experiences of leaders, rather than the availability of technology. The Egyptian case revealed that similarly to protests before the digital age, leaders needed to work over long periods, building their networks with different groups and preparing their mobilisation strategies in advance for the sustainability of their protests. It is also seen that when a performance was familiar, leaders could also more easily co-ordinate resources such as ICTs around a common goal and adapt their activities according to the changing structures.

The analysis also showed that organisations were not completely absent in the organisation of the Egyptian protests and that they provided the activists with the ability to adjust course. For instance, operating as non-legitimised actors, the long-established political organisations such as the Muslim Brotherhood, NGOs and human right groups co-ordinated social networks and marshalled resources, such as logistical and technological support. Thanks to their past experiences and resources, these NGOs and human right groups were able to engage in the tactical and decision-making manoeuvres necessary to sustain the movements. Thus, the Egyptian protests were not personalised and loosely co-ordinated actions, corresponding to the crowd-enabled movement definition of Bennett and Segerberg (2013). They rather reflected the features of organisationally enabled movements, an action type that lies between the organisationally brokered and crowd-enabled types (Bennett and Segerberg 2013: 48). On the other hand, the Syrian protests better represent the features of a crowd-enabled movement, with no formal organisation or co-ordination of action. They emerged from the local interaction of numerous individual actors. Yet, unlike Bennett and Segerberg's (2013) definition of crowd-enabled movements, technology was not the prominent agent of mobilisation in the Syrian uprising.

Finally, the findings showed that the sustainability of the protests was dependent upon the resources available to the protesters: (1) a political public with a vibrant civil society was crucial to leading the protests, with organisations such as NGOs and human right groups co-ordinating logistical and technological support by activating their social networks. (2) Division amongst the elites—particularly military elites and business elites—was also crucial to reducing the state's power and making it vulnerable to collapse. (3) Support from different segments of society was another significant factor that affected the economic power of the state and had the power to essentially bring economic life in the country to a standstill. (4) It was also seen that, in countries where the regime actively used the media to generate a consistent narrative, as in the case of Syria, the support of protesters from different segments of society declined. Moreover, with the ban of broadcasting channels and partial restriction of Internet access, it was apparent that the protesters struggled to remain in contact both with each other and also the world. Thereby, the hybrid media activities of the regime were also key to affecting the success of the protests. All four of these factors were also found by Howard and Hussein (2013), who conducted a fuzzy set analysis of 22 Arab uprisings and questioned the causal factors that underpin the sustainability of protests. Despite the differing methods used in both this research and theirs, these four factors were determined to be sufficient conditions for the success and sustainability of the protests. Additionally, I also found that (5) the level of state repression was a significant factor that triggered public anger. In the presence of a vibrant civil society, as in the case of Egypt, repression helped bring success to the activists. However, it was seen that in the absence of a civil society, this repression might also trigger the radicalisation of small groups, as was the case in Syria. Finally, (6) the sixth factor concerned the diffusion of power in the state. It was noted that when power is centralised in one person, as in the case of Egypt, a swift transition without dismantling the regime was possible. This helped the army to swiftly replace the president and accelerate the process of revolution. More importantly, the data showed that the sustainability of protests not only depends on these six resources but also (7) to the ability of leaders to use them.

In the conclusion chapter, alongside the main arguments of this book, I discuss why the long-term change that the 25 January protesters hoped to see in Egypt has still not arrived in the country and explain the weakness of digitally supported protests in bringing the social change at authoritarian states.

NOTES

1. Interview with Abdelrahman Mansour, the administrator of the Kullena Khaled Said page, London, 2015.
2. Interview with Mahmoud (pseudonym), the administrator of the Syrian revolution page, London, 2013.
3. Interview with Abdelrahman Mansour, the administrator of the Kullena Khaled Said page, London, 2015.
4. Interview with Mahmoud (pseudonym), the administrator of the Syrian revolution page, London, 2013.
5. Interview with Abdelrahman Mansour, the administrator of the Kullena Khaled Said page, London, 2015.
6. Interview with Sami (pseudonym), Egyptian activist, Cairo, 2014.
7. Interview with Abdelrahman Mansour, one of the Kullena Khaled Said administrators, London, 2015.
8. Interview with Mona (pseudonym), Egyptian activist, Cairo, 2014.
9. Interview with Mona (pseudonym), Egyptian activist, Cairo, 2014.
10. Interview with Osama, Syrian activist, Istanbul, 2013.
11. Interview with Abed Alsarraj, Syrian activist and journalist, Istanbul, 2013.
12. Interview with Sherif Azer, Egyptian activist, London, 2014.
13. Interview with Mona (pseudonym), Egyptian activist, Cairo, 2014.
14. Interview with Esraa Abdel Fattah, Egyptian blogger and activist, Cairo, 2014.
15. Interview with Mona (pseudonym), Egyptian activist, Cairo, 2014.
16. Interview with John Albert, Egyptian activist, Alexandria, 2014
17. Interview with Mizyan Altawil, Syrian activist, Istanbul, 2013.
18. Interview with Enrico Angelis, researcher, Cairo, 2014.
19. Interview with Mizyan Altawil, Istanbul, 2013.
20. Skype interview with Caroline, 2013.
21. Interview with the employees of Syrian National Coalition, 2013.
22. Interview with John Albert, Egyptian activist, Alexandria, 2014.
23. Interview with Mahmoud (pseudonym), the administrator of the Syrian revolution page, London, 2013.
24. Interview with Caroline, Syrian activist, Istanbul, 2013.
25. Skype interview with Hediye Levent, journalist, 2013.
26. Interview with Osama, Syrian activist, Istanbul, 2013.

REFERENCES

Abdelrahman, M. (2014). *Egypt's long revolution: Protest movements and uprisings*. New York, NY: Routledge.

Agarwal, S. D., Bennett, L., Johnson, C., & Walker, S. (2014). A model of crowd enabled organization: Theory and methods for understanding the role of Twitter in the occupy protests. *International Journal of Communication, 8*, 646–672.

Al Hendi, A. (2012, October). The kingdom of silence and humiliation. *Foreign Policy*. Retrieved from http://foreignpolicy.com/2012/10/16/the-kingdom-of-silence-and-humiliation/.

Aslan, B. (2015). The mobilisation process of Syria's activists: The symbiotic relationship between the use of information and communication technologies and the political culture. *International Journal of Communication, 9*, 2507–2525.

Athamneh, W., & Sayej, C. M. (2013). Engaging the authoritarian state: Voices of protest in Syria. *Journal of Arabic and Islamic Studies, 13*, 169–190.

Azimi, N. (2005, September 1). *Egypt's youth have had enough*. Open Democracy. Retrieved from https://www.opendemocracy.net/democracy-protest/enough_2794.jsp.

BBC News. (2011, April 19). *Syria witness: 'Unprovoked shooting' at protest in Homs*. Retrieved from https://shorturl.at/iJKNT.

Bennett, W. L., & Segerberg, A. (2013). *The logic of connective action*. Cambridge, UK: Cambridge University Press.

Bennett, W. L., Segerberg, A., & Walker, S. (2014). Organization in the crowd: Peer production in large-scale networked protests. *Information, Communication & Society, 17*(2), 232–260. https://doi.org/10.1080/1369118X.2013.870379.

Chadwick, A. (2013). *The hybrid media system: Politics and power*. New York, NY: Oxford University Press.

Cole, J. (2014). *The new Arabs: How the millennial generation is changing the Middle East*. New York, NY: Simon & Schuster.

Ghonim, W. (2012). *Revolution 2.0*. Boston, MA: Houghton Mifflin Harcourt.

Goodwin, J., & Pfaff, S. (2001). Emotion work in high-risk social movements: Managing fear in the U.S. and East German civil rights movements. In J. Goodwin, J. M. Jasper, & F. Polletta (Eds.), *Passionate politics: Emotions and social movements* (pp. 282–302). Chicago, IL: University of Chicago Press.

Gunning, J., & Baron, I. Z. (2013). *Why occupy a square: People, protests and movements in Egyptian revolution*. New York, NY: Oxford University Press.

Haddad, B. (2012). Syria, the Arab uprisings, and the political economy of authoritarian resilience. *Interface, 4*(1), 113–130.

Hanisch, C. (2001). Struggles over leadership in the women's liberation movement. In C. Barker, A. Johnson, & M. Lavalette (Eds.), *Leadership and social movements*. Manchester, UK: Manchester University Press.

Harding, L., & Arthur, C. (2013, April 30). Syrian electronic army: Assad's cyber warriors. *The Guardian*. Retrieved from https://bit.ly/2SQofVN.

Herrera, L. (2014). *Revolution in the age of social media*. London, UK: Verso.

Howard, P., & Hussain, M. (2013). *Democracy's fourth wave? Digital media and the Arab Spring*. New York, NY: Oxford University Press.

Human Rights Watch. (2009, November 26). *Group denial*. Human Rights Watch. Retrieved from https://www.hrw.org/report/2009/11/26/group-denial/repression-kurdish-political-and-cultural-rights-syria#.

Jenkins, J. C. (1983). Resource mobilisation theory and the study of social movements. *Annual Review of Sociology, 9*, 527–553.

The Kefaya Movement. (2008). *RAND National Defense Research Institute*. Retrieved from http://www.rand.org/content/dam/rand/pubs/monographs/2008/RAND_MG778.pdf.

Landis, J. (2011, March 31). *Speech to the Syrian parliament by president Bashar Al-Assad: Wednesday, March 30, 2011* [Web log post]. Retrieved from http://www.joshualandis.com/blog/speech-to-the-syrian-parliament-by-president-bashar-al-assad-wednesday-march-30-2011/.

Leenders, R. (2012, December). Collective action and mobilization in Dar'a: An anatomy of the onset of Syria's popular uprising. *Mobilization, 17*(4), 419–434.

Lesch, D. W. (2012). *Syria: The fall of the house of Assad*. Cornwall, UK: TJ International.

Levinson, C., & Coker, M. (2011, February 11). The secret rally that sparked an uprising. *The Wall Street Journal*. Retrieved from http://www.wsj.com/articles/SB10001424052748704132204576135882356532702.

Local Coordination Committees of Syria. (2012, December 20). *Carnegie Endowment for International Peace*. Retrieved from http://carnegieendowment.org/syriaincrisis/?fa=50426&reloadFlag=1.

McCammon, H. J., Newman, H. D., Courtney, S. M., & Terrell, T. M. (2007). Movement framing and discursive opportunity structures: The political successes of the U.S. women's jury movements. *American Sociological Review, 72*(5), 725–749.

Najm, A. A. (2011, October 1). Syria's coordination committees: A brief history. *Al-Akhbar*. Retrieved from http://english.al-akhbar.com/node/764.

Naggar, M. (2011, May 9). *Media at war in Syria*. Doha Centre for Media Freedom. Retrieved from http://www.dc4mf.org/en/content/media-war-syria.

Oberschall, A. (1973). *Social conflict and social movements*. Essex, UK: Pearson Education.

Papacharissi, Z., & Oliveira, M. (2012). Affective news and networked publics: The rhythms of news storytelling in Egypt. *Journal of Communication, 62*(2), 266–282. https://doi.org/10.1111/j.1460-2466.2012.01630.

Radsch, C. (2008, September). Core to commonplace: The evolution of Egypt's blogosphere. *Arab, Media & Society, 6*. Retrieved from http://www.arabmediasociety.com/?article=692.

Rogin, J. (2011, February 2). White house failing to consent Mubarak to start transition now. *Foreign Policy.* Retrieved from http://foreignpolicy.com/2011/02/02/white-house-failing-to-convince-mubarak-to-start-transition-now/.

Salem, S. (2013, September 6). *The Egyptian military and the 2011 revolution.* Jadaliyya. Retrieved from http://www.jadaliyya.com/pages/index/14023/the-egyptian-military-and-the-2011-revolution.

Sha'ban, A. B. (2006). *The butterfly effect: Kefaya—Past and present.* Cairo, Egypt: Kefaya Printings.

Shehata, D. (2011, May/June). The fall of the pharaoh: How Hosni Mubarak's reign came to an end. *Foreign Affairs.* Retrieved from https://www.foreignaffairs.com/articles/north-africa/2011-04-14/fall-pharaoh.

Stacher, J. (2012). *Adaptable autocrats: Regime power in Egypt and Syria.* Stanford, CA: Stanford University Press.

Tarrow, S. (2011). *Power in movement: Social movements and contentious politics.* New York, NY: Cambridge University Press.

Tilly, C. (1985). Models and realities of popular collective action. *Social Research, 52*(4), 717–747.

Tilly, C. (2006). *Regimes and repertoires.* Chicago, IL: University of Chicago Press.

Tufekci, Z. (2017). *Twitter and tear gas: The power and fragility of networked protest.* New Haven, CT: Yale University Press.

Wall, M., & Zahed, S. (2011). "I'll be waiting for you guys": A YouTube call to action in the Egyptian revolution. *International Journal of Communication, 5*, 1333–1343.

Zisser, E. (2001). Does Bashar al-Assad rule Syria? *Middle East Quarterly, 10*(1), 15–23. Retrieved from http://www.meforum.org/517/does-bashar-al-assad-rule-syria.

CHAPTER 8

Conclusion

On 28 April 2014, the court ordered the disbanding of the April 6 Youth Movement and the freezing of its activities. The ban came as a result of the Protest Law issued in November 2013 that empowered the Egyptian authorities to ban and dissolve most public demonstrations ("We will not be silenced" 2014). On that day, I was waiting for the prominent Egyptian blogger, Wael Abbas, in a café in new Cairo. When he finally arrived, his frustration and disappointment were apparent. Wael[1] mentioned that he was following the news about the April 6 movement and had been contacting lawyers throughout the morning. I asked him whether he was positive about the future of the revolution, to which he replied:

> Not really, all I can do is continue working but I cannot guarantee anything. People got killed for the weirdest reasons. I can get killed on the street by police or by some angry person who saw me on TV and thinks that I'm a traitor. Nothing is guaranteed here. There is a complete absence of law in the street. The police and army are only protecting the regime, but people are left to clash with each other on the streets. Nobody is protected. Carjacking and activist abduction are on the rise.

Wael's frustration was indicative of the spirit prevalent amongst many of my interviewees in 2014. Meeting these expectations, in 2018, the Sisi government, which had ascended to power in July 2013, ramped up

B. Aslan Ozgul, *Leading Protests in the Digital Age,*
Palgrave Studies in Young People and Politics,
https://doi.org/10.1007/978-3-030-25450-6_8

its mass arrest campaign. Alongside Wael, many activists, human rights workers and lawyers who provided humanitarian and legal support to the families of political detainees were arrested (Human Rights Watch 2018a). The repression of dissent was presaged the revival of authoritarian rule and the unchanged social order in Egypt. Indeed, few years ago, the emotions of hope and joy accompanied the word "revolution" in Egypt. The rise of crowds in the absence of conventional organisations and their success in the overthrow of Mubarak was praised by many social movement and communication scholars, who drew a connection between technology diffusion, the use of social media and political change in 2011 (Castells 2012; Howard and Parks 2012: 360).

As research increased, scholars began to criticise this reductionism in communication. Media technologies were conceptualised as mere instruments in the hands of activists and highlight the power of the people using these new technologies rather than the power of the technology itself (Aouragh and Alexander 2011; Eltantawy and Wiest 2011; Trere 2019). They also found that not only the Internet, but also traditional communication methods, such as face-to-face, played significant roles in organising and mobilising the Arab uprisings (Aouragh and Alexander 2011).

Today, the heated debate continues with regard to the complex, multifaceted role of communication technologies within protest movements, as well as their capabilities and strengths. As was noted in different chapters in this book, Bennett and Segerberg's (2013: 194–195) ground-breaking book brought a new point of view to this debate. Analysing the Occupy Wall Street protests, they demonstrated that communication technologies are not only instruments in the hands of people but became organising agents and changed the dynamic of action. The authors argued that one can identify three types of actions in the digital age. Within the first action, which shows the dynamics of conventional collective action, digital media do not significantly alter the outcomes, but in the purest form of connective action, the crowd-enabled type, ICTs become organising agents and change the dynamic of action, replacing the need for collective action framing, resource mobilisation, strong leadership and collective identity (Trere 2019).

Bennett and Segerberg's (2013) argument was important for pointing out the peer production and sharing culture of the Internet and its influence on the organisation of digitally supported protests. However, as Trere (2019) has emphasised, conceiving the new media technologies as mere instruments or overemphasising their organisational role are fallacious practices in the current literature. While Bennett and Segerberg's (2013) arguments on ICTs are undoubtedly relevant and push us to consider the new organisational dynamics of today's movements, they pay excessive attention to the dynamics of public, external communication on social media platforms such as Twitter, at the expense of internal communication and organisation in today's movements. To understand the actual capabilities of digitally supported actions, we must take a more holistic approach that explores both the external and internal dynamics of a movement and considers the daily communicative exchanges through which activists organise, create and nurture collective identities (Trere 2019).

To this end, I commenced this research by focusing on the communicative origins of digitally supported protests. Analysing who created the first message and attempted to mobilise the public in the online and/or offline spheres helped me to show how these first activists mapped, understood and navigated ICTs. To understand the affordances of ICTs in the organisation of protests, I analysed the mobilisation, co-ordination and internationalisation attempts of activists and how they used ICTs at these stages. While engaged in this, I also analysed the tension between states and activists to identify and delineate the necessary mechanisms and dynamics for the sustainability of today's protests.

Below, I retrace the central arguments of this book, which were initially defined and addressed separately in the two cases analysed, i.e. the 2011 Egyptian and Syrian protests. I also note some of the limitations of my research and address future questions that ought to be asked to fully understand today's protests. In the final section, I discuss the challenges and opportunities offered by digitally supported protests.

The Three Styles of Leadership

As Tufekci (2017: 270) claims, "the participatory structure is the very reason many people join protests in the first place". However, behind this non-hierarchal, participatory capacity, we can identify key figures who mobilised and co-ordinated the protests. The new media landscape does not generate leaderless movements, but rather helps the leaders to operate anonymously and invisibly jockey for position. Analysing the preparation of the Egyptian and Syrian uprisings and asking how the first protest message appeared in the respective countries enabled this study to capture the specific role of three styles of leaders. Conforming to the nature of crowd-enabled movements, the Egyptian and Syrian movements appeared to be leaderless, with no spokesperson or institutional leadership. However, the findings of this book revealed that in both cases, the administrators of social media pages were the first actors who tried to mobilise the public. In line with the findings of Gerbaudo (2012: 157), who analysed the digitally supported protests in Mexico, Egypt and the USA, they demonstrated soft and emotional leadership skills, as well as gatekeeping capacities, determining the discourse that would be adapted to the page (Gerbaudo 2012: 157). By conducting interviews with the administrators of the two main Facebook pages co-ordinating protests in the two countries, this study revealed that unlike the leaders of professional social movement organisations, as defined by McCarthy and Zald (1977), these "soft leaders" did not expose themselves on television to form the impression of widespread activity and grievance. Rather, they preferred to remain anonymous and operated behind the scenes. The anonymity the Internet provided was important for two reasons—firstly, it was secure against an oppressive response by the state and, secondly, it attracted the support of large segments of society. In Chapter 5, the Syrian case showed that when the Facebook page specified a spokesperson with a Muslim Brotherhood background, the support to the page fell. This revealed that to unify the Internet users from different backgrounds, the administrators needed a shared grievance or iconic figure with whom users could identify, rather than a charismatic leader with a specific ideology. Often acting anonymously, a soft leader frequently became the first person to highlight the corruption and repression they experienced under the regime, gave voice to the problems of citizens and assisted in melting away the divisions amongst the followers of the page. The non-hierarchal structure of the social media pages also permitted these leaders to invite their followers to participate in the decision-making process and assist in the construction of a collective identity amongst page followers. However, Chapter 7 showed that as having to consult large numbers in order to decide on every act of the movement would create conflict and slow down the activists' activities, soft leaders do not always adhere to the collective leadership norm of networked protests emphasised by Bennett and Segerberg (2013). As in the case of Egypt, few figures may determine the discourse and content shared on the page.

The efforts of the Syrian soft leaders to get visibility for the page also showed that broadcast media remains crucial for the prominence of soft leaders; they are the actors who brought fame to videos uploaded on YouTube or to social media pages. The research also further developed the soft leadership argument of Gerbaudo by introducing two more leader groups that appeared in digitally supported movements. What were those two groups?

First, both the Syrian and Egyptian uprisings showed that the experience of activists on the ground was as significant as the role of the soft leaders in encouraging societies which had barriers to overcome between willingness to participate and effective participation. Acting as strategic decision makers on the ground, the experienced activists effectively managed divisions of labour, set movements' goals, mobilised the poor segment of society and led the protests offline. In Chapter 4, it was seen that by dividing into small mobilisation groups, the experienced activists of Egypt communicated face-to-face with the public in different neighbourhoods and distributed protest information using leaflets and brochures. Similarly, the experienced activists in Syria also used snowballing mobilisation tactics. However, as they had not previously formed social networks with different groups, they merely galvanised their friends or relatives. Certainly, these experienced activists also used the Internet to inform the public or discuss the logistical and strategic issues within their closed circle. However, as Mona explained in Chapter 3, they did not play a significant role on the Internet as an administrator of a social media page where they would create unity and togetherness amongst the online followership. On the other hand, the soft leaders of both Egypt and Syria were people who lived abroad for many years and began to mobilise support for the protests from distant countries. As they did not have the solid connections and knowledge on the ground that experienced activists did, they left the co-ordination of the protests themselves to experienced activists.

The case of Egypt also demonstrated the rise of a new kind of leadership in today's movements, which I termed hybrid leaders. As was seen in Chapter 3, some of the administrators of social media pages, like Esraa Abdel Fattah or the administrators of Kullena Khaled Said page, who were familiar with the essential tactics of information dissemination on the Internet, became the communicators both on the ground and Internet. They reflected the function of leaders that Morris and Staggenborg (2002) cite: they acted as "mobilisers", inspiring participants and also as invisible "articulators". Despite their anonymity on social media, they could appeal to different segments of society, drawing on followers' social and economic problems or gender identifications. They thereby touched upon societies' cultures and linked the aspirations of different Egyptian classes to those enshrined in Egyptian culture (Morris and Staggenborg 2002).

The hybrid leaders also framed grievances on social media through the reporting of news from the ground. In Chapter 3, it was shown that the administrators of the Kullena Khaled Said page went to speak with the family of missing people and reported them on social media. The difference from the soft leaders is that they themselves went to areas with problems and reported the issues from first hand using ICTs. They also mobilised the public on the ground by distributing leaflets, brochures and face-to-face communication. Hence, the hybrid leaders blended together the pre-existing protesting styles of experienced activists with new media logics of soft leaders. While they operated on the Internet, they referred to the sensibilities of urban middle-class segments that wanted liberal politics and freedom of expression. On the other hand, while communicating with the poor and workers on the ground, the hybrid leaders referred to the demands of this segment for bread and improved standards of living (Cole 2014: 269). Mobilising and orchestrating these groups with different demands from different segments of society required an intellectual background. The hybrid leaders were mostly from the young, educated middle-class segment of society and hence were accomplished in the manipulation of language and symbols in both spheres. These functions of leaders are cited as crucial for the success of movements by social movement theorists such as Morris and Staggenborg (2002). In Chapter 7, it was seen that the lack of this intellectual leadership in Syria obstructed the processes of coalition-building and the channelling of emotions amongst elites.

In my analysis, I focused on the protest experiences of the leaders in the 2011 Egyptian and Syrian protests. Much more research is needed to test whether the three kinds of leaders, emerging from the sample of this study, were present in the digitally supported protests around the world. Analysing what the different mobilisation tactics undertaken by these leaders were, and how their capacities were affected by structural factors, would be illuminating.

The Symbiotic Relationship Between Repertoire and Use of ICTs

An important point about digitally supported protests is that they are usually characterised as being formed by loosely tied networks without a collective identity. Yet, as highlighted throughout this book, the Egyptian soft and hybrid leaders have not only used ICTs to gather people around a common aim and organise large rallies, but also benefitted from these new technologies in creating trust relations and a feeling of belonging amongst the protest participants. My starting point in this research was to discover why the Syrian activists could not benefit from ICTs in the same way as their Egyptian counterparts did. Discovering the political, social and cultural contexts in Egypt and Syria in Chapter 2 was key to understanding which of the factors most affected the use of ICTs during the protests. It showed that unlike the Egyptian youth groups who shared their personal views with different ideological groups in the Egyptian blogosphere, an online community was absent in Syria before the 2011 protests (Angelis 2011). The repertoires entailing social ties, past performances and identities which were necessary for collective claim-making were still weak.

The interviews with the Syrian activists showed that mobilising a public through the Internet—a public who lack established repertoires—demands time and a cause with which page followers can identify themselves. As was seen in Chapter 5, fear of a potentially oppressive response from the regime kept Syrians from immediately responding to the call for protests. The repression level in Syria, which had emerged due to the national political structure, had a significant impact on this fear and caused to the lack of protests before 2011. Those who endeavoured to oppose the regime met with the heavy repression unleashed by Syria's security forces (Van Dam 1996). The tense political atmosphere following independence further precipitated the reluctance of Syrian citizens to participate in politics. The sectarian discourse of the regime and its attempts to create mechanisms of control by semantic practices further alienated the public from political life (Weeden 1999). Similarly to his father's reign, during the Bashar al-Assad presidency an effective cult of personality around the president was constructed through nationalist rhetoric and symbols (Magout 2012). The orchestrated spectacles, for instance, were occasions at which participants were disciplined and organised for the physical performance of ritual gestures (Weeden 1999:

19). The spectacles reflected instantiations of power. They filled public spaces with political ideas and the regime's iconography. Although the behaviour of the public did not always reflect their true values and beliefs, obeying these norms made them feel safe. Thus, the cult created a powerfully yet ambiguous mechanism of social control, leading to public dissimulation (Weeden 1999: 70). The notion of a revolt against the regime was unimaginable, frightening and insecure.

To break this fear, a preparation time in which the administrators created trust relations amongst the participants and established a notion of causality and belonging was important in fostering the digitally supported protests. As Sidney Tarrow (2011) argues, "leaders can create a social movement only when they tap into and expand deep rooted feelings of solidarity or identity" (Tarrow 2011: 11). The administrators of "the Syrian Revolution 2011" Facebook page, who were the first actors that tried to mobilise the Syrian public against the regime, could not create this trust relation in a mere two weeks. Fear not only affected the public's reactions to protest calls, but also the performances and repertoires of the organisers of the movements. After all, mounting a demonstration depended on the past performances and repertoires of the actors involved (Tilly 2006: 35). As the protesters had long lived with fear of coercion structures, they lacked experience of online activism against the state. The administrators opted for an ill-conceived narrative, spokesperson, profile photo and protest date that coincided with the Muslim Brotherhood uprising of 1982. Moreover, the administrators had not yet discovered the importance of iconic figures and specific events in activating Internet users. They had difficulty activating people's emotions in the absence of an iconic figure or specific event that carried profound emotional power, as had the death of Khaled Said in Egypt and the torture of children in Daraa. Consequently, they could not mitigate fear amongst Syrians and motivate them to participate in the protests. In Chapter 7, the case of the Kullena Khaled Said page of Egypt showed that digital networks can enable the formation of a collective identity amongst middle-class followers of the page around a figure, but that this process requires time, dedication and experience. The administrators of SRFP had neither this time nor the experience to create a collective identity amongst the page followers.

The reason why the Internet served merely as a tool for exchanging information rather than the Syrian protesters' main organisational hubs related to this fear and lack of protest repertoires. Absent established social networks, the Syrian protesters did not trust each other online. Fearing arrest by the Shabia, they preferred to follow the conventional culture of communication in Syria (Rinke and Roder 2011: 1275). They benefitted from oral communication to create trust with others. The experienced activists who led the protests used the mosques as the new protest sites, thereby eliminating the organisational and logistical issues that a social movement organisation would have encountered. People gathered in the mosques without the risk of showing up alone in a protest area; there, interviewees report, they felt deep-rooted feelings of solidarity. Online communication could not produce large mobilisations due to the protesters' security precautions even though information did spread through personalised networks. Rather, protesters opted for one-to-one mobilisation tactics rather than "many-to-many" when the protests first began. The Syrian case was a good example, demonstrating that the power of today's protests has changed in accordance with the context in which the movements emerge and the leaders' abilities to use the resources available to them.

But despite the vulgarity of the political context, the Internet activities of the protesters expand as their repertoire of contention develops. The co-ordination of local committees, which emerged in different cities, and their internationalisation attempts, as described in Chapter 6, created a symbiotic relationship between the repertoire of contention and use of ICTs. The more often Syrian protesters participated in actions and developed their repertoire, the more their online and offline activities became synchronised and organised. They bypassed the media blackout imposed by the regime and the structural restrictions through the effective use of satellite phones and the Internet.

The Relevance of Resource Mobilisation Theory to Today's Digitally Supported Movements

When the protests broke out in Tunisia, the images of revolution on the broadcasting channels and ICTs brought sorrow and hope to fellow protesters in Arab countries and created shared grievances (Castells 2012; Howard and Hussein 2013). The administrators of the Facebook pages—Kullena Khaled Said in Egypt and the Syrian Revolution 2011 in Syria—assembled dispersed people and invited them to take part. They created a "choreography of assembly", to use Gerbaudo's (2012) term.

The images and personalised messages of protest spread virally across the social networks of Kullena Khaled Said's followers. Observing mobilisation that spilled outside the bounds of formal organisations and group identity slogans led to many big data researchers concluding that ICTs present new political opportunities in today's movements that help organisers activate the public without resources that were once crucial for the emergence and sustainability of collective actions.

However, the research which triangulate quantitative research on big data with qualitative ethnographic work, showed that while the Facebook pages were a stitching mechanism for the first stage of the uprising, other communication resources such as face-to-face communication and the distribution of leaflets and brochures were as important as the social media for the organisation and mobilisation processes of the protests. Conducting interviews with key activists both on the ground and on the Internet and tracing the actual course of events simultaneously with the activists' repertoire of contention, beliefs and attitudes, I tried to develop these studies and demonstrate how different means of communication are related with different steps in mobilisation processes. For instance, in Chapter 5, it was seen that ICTs were one of the first means of communication in the Syrian mobilisation process. However, we had also seen the continuing importance of cafes and the rising significance of mosques for the first stages of the mobilisation and co-ordination processes.

The findings also revealed that the mobilisation, co-ordination and sustainability of protests still depend on the capabilities of leaders. Today's digitally supported leaders also operate within structures and their capabilities both influence and are influenced by a movement's organisation and the political context it operates in (Morris and Staggenborg 2002: 174). In Egypt, for instance, the suitable political environment helped the development of leaders with: (1) an embedded and long-standing protest history; and (2) multiparty presidential elections. These multiparty presidential elections of Egypt gave way to the appearance of political figures like ElBaradei who would become the hope for change before 2011 and spur the youth to acquire experience in online and offline spheres. On the Internet, young people would already establish social networks with internal and external groups before the 25 January protests and this would enable them to orchestrate the 25 January protests in synergy with these groups. It was also on these platforms that bloggers and the administrators of social media pages practised citizen journalism and stoked the political consciousness of the middle-class segment. Acquiring experience of online platforms helped these administrators use the technology most effectively during the protests and create the necessary online networks.

The capabilities of the leaders also depended on social networks established on the ground. The Egyptian administrators and the urban activists of Cairo created social networks with different segments of society on the ground over the years. These networks helped them to frame grievances, devise strategies and tactics for different segments of society and unify them around the same aim to ultimately topple the long-standing dictator, Hosni Mubarak. In Egypt, for instance, the administrators of the April 6 movement Facebook page contacted the workers in El Mahalla El-Kubra to create social networks between Cairo and El Mahalla El-Kubra before the 2011 protests emerged. The administrators learned from the workers of El Mahalla El-Kubra how to mobilise the poor segment of society. During the protests, this connection also allowed them to quickly recognise and act on opportunities in different cities. The rural leaders of Syria, on the other hand, lacking past protest experience and the same intellectual background, failed to connect the urban cities with rural areas to unify protesters around secular, non-ideological aims and prevent extremism from emerging. They subsequently lost the support of urban elites over the course of the protests (Landis 2011).

The sustainability of digitally supported protests not only depended on the capabilities and background of leaders. Like the traditional social movements, sustainability still varied according to the resources controlled by the state and those pooled to the movement by outside organisations (Jenkins 1983: 528). The resource mobilisation approach of this book emphasises the resources beyond money, labour, time and funding from organisations such as churches or businesses that have been cited by resource mobilisation theorists McCarthy and Zald (1977). Chapter 4 showed that in the digital age, resources might be transferred to a movement from (3) outside organisations such as the NGOs, human right organisations, political organisations and (4) an active opposition media that covered torture and repression in the country. For instance, the plurivocal media and civil society in Egypt provided activists with the opportunity to reach different segments of the population with or without Internet access. While the opposition media was significant in reprinting and diffusing the oppositional voices, human rights organisations played a role in picking up and retransmitting them during the protests. The established political organisations such as the Muslim Brotherhood and NGOs were also significant in providing technical, logistical and networking support to the protesters on the ground. Sustainable and effective action, which can respond to external changes, still requires resources such as friendship and activist networks, tactical repertoires and the resources that the organisations such as Muslim Brotherhood offer.

In Syria, on the other hand, despite the efforts of the civil society movement at the beginning of 2000, the influence of the state was still dominant in all institutions, including the media and NGOs (Hinnebusch 2012: 103). The Syrian state also used hybrid media effectively during the protests. It disseminated messages that delegitimised the protests and protesters through social media with an electronic army and used the national mainstream media to enforce this discourse. The state also used its capacity to substantially restrict Internet access in cities under siege by the army. With the entrance of international journalists into the field also banned, protesters found it difficult to get their voices heard in the first months of the protests. Lacking the support of a strong civil society and international journalists, Syrian protesters relied on the technological support of the diaspora to reach the outside world. Although this support enabled the Syrian activists to be heard, it could not help their internal co-ordination. The resources of the Syrian state and its capacity to use these were therefore strategic factors that differentiated the phases of the Egyptian and Syrian protests.

There is no doubt that ICTs offer new opportunities to today's leaders but (5) the diffusion of power in political structure within which a movement emerges remains important in the digital age. The two cases represent two neoliberal authoritarian regimes that utterly differ from one another (Stacher 2012: 157). For instance, the centralised political structure facilitated a change of power in Egypt. Thanks to this centralised system, the army, which was cynical towards the new cluster of elites affiliated with Gamal Mubarak, could easily change the power dynamics in Egypt. On the other hand, the power in the Syrian regime is diffused amongst three institutions, namely the Ba'ath Party, president and army. These institutions form a cross-sectarian ruling coalition (Stacher 2012: 15). The sectarian tendencies of the Syrian state created a dependent relationship between the two power structures of the army and president and helped the latter retain the support of the army during the protests.

As the army backs the state, (6) state repression in Syria escalated. The high degree of repression, which also resulted in the arrest of the few experienced activists who could guide the protesters on the basis of their fruitful experience, would serve the interests of the Syrian state, as it led to the appearance of armed opposition groups after six months of protests. The sectarian tendency of the government had also encouraged an Islamist resurgence, which would curb the support of urban elites for the Syrian protests in subsequent stages of the uprising. The differences in the six resources cited above are still determinant in defining the path that a movement will take in both the digital and offline spheres.

Although this book discusses the different sets of factors that affect the sustainability of digitally supported movements, its aim was not to explore each of these factors in detail. This research focused in particular on the organisational structure of the digitally supported protests and the impact of this structure and the use of ICTs on the efficacy of the protests. The data on the political structure of the regime and its impacts on the protests are drawn from the works of scholars such as Stacher (2012), who rigorously analysed and compared the distribution of power in both regimes and the impacts of this on the protests. The book also refers to the sectarian tendency of the Syrian state but it does not explore in detail the effects of this tendency on the protests. Much more detailed research is needed to consider the extent to which the sectarian structure and discourse of the Syrian state helped its own sustainability during and after the Syrian protests. Moreover, since this research focuses on peaceful protests within two countries, it does not discuss the role of the international community. International society acquired a particular role after the first six months of the protests in Syria. The analysis of international society requires focused research that goes beyond the objectives of this book.

THE CHALLENGES AND OPPORTUNITIES OF DIGITALLY SUPPORTED MOVEMENTS

In Chapter 6, it was seen that the Syrian state's effective use of repression and a hybrid media system produced divisions amongst the Syrian protesters. After six months, some eventually took to arms to prevent violent repression against opposition groups. This led to a battle between the opposition and forces loyal to Bashar al-Assad, drawing in neighbouring countries, world powers and extremist groups, including the Islamic State. As of March 2018, the estimated death toll in Syria was more than 500,000 (Human Rights Watch 2018b). The post-Mubarak period in Egypt saw the rise and fall of the Muslim Brotherhood government, resulting in the military ousting of the first elected Egyptian president, Mohammed Morsi, in 2013. Since then, Egypt, in turmoil, has started another era of military rule. During all these stages, the old state, with its competing institutional power centres, managed to persist (El-Sherif 2014). All of these developments led many observers to question whether the 2011 events amounted to a revolution in Arab countries. As Cole (2014: 268) argues, Egypt has long been ruled by a family cartel that concentrated tremendous wealth in the hands of the Mubarak family and the elite cluster around the family that became billionaires through government contracts. The removal of Mubarak led to a vast redistribution of wealth (Cole 2014: 268). Hence, one could observe a political revolution in Egypt that was "a rapid and systemic political change brought by a social movement" (Cole 2014: 267). However, the developments in both countries showed that although digitally supported movements brought political change rapidly, they are not capable of bringing simultaneous social change. Changing the presidents in these states often proved to be the easy part of the revolution, the main challenge was to revolutionise the deeper social cleavages and structures (Stacher 2012: 3). Social revolutions were indeed embedded in a long-term process. As the Middle East expert Professor Asef Bayat argues, social change in an authoritarian state can only arrive if every social group generates change in society through active citizenship (Bayat 2013: 313). In Arab societies, which long lived under the dominance of states that could extend their control well beyond formal institutions, active citizenship was a notion to be instilled. It was notable that the rise of social media and satellite broadcasters started to change the long-reigning fear in Arab societies. However, the absence of active political citizenship could still be particularly felt in Syria, where the fear of the regime made members of the civil society reluctant to participate in political life.

In Egypt, the protests that continued in the 2000s created new groups, networks and cultural products. Yet, when the street protests finished and left routine politics in their place, there was an absence of representation and delegation for the revolutionaries. In line with the argument of Bayat, the Egyptian interviewees of this research claimed that the problem arose due to the reluctance of the protesters to participate in political life. One of the administrators of KKSFP, Abdelrahman Mansour, said,[2] for instance:

> As revolutionary youth, our main problem was that after the revolution, none of us wanted to gain power or become a politician. All of us wanted to remain revolutionaries. That does not work; you are a revolutionary and then you need to transition into a politician to reach your political goals. We failed in that.

Apart from a few revolutionaries, such as the blogger Mahmoud Salem (alias: Sand Monkey) who would run for the Egyptian parliament in 2011, the organisers' reluctance shows that the true reform of authoritarian states requires laborious struggle. Protests needed to be more than discursive struggles in which the revolutionaries consolidated their institutional foundations within the fabric of society (Bayat 2013: 313). In the absence of new figures in the parliament, the electoral and institutional system, as well as police and army had remained more or less intact.

The regimes have also gradually developed themselves in controlling the new public sphere and its digital ecology. Chapter 6 showed that the Syrian cyber army often made rhetorical attacks and confused the silent segment in Syria who were reluctant to join in the protests. Today, the counter-movement and new surveillance techniques of governments are the real challenges, cutting the connectivity between the dissidents and the public and leading to intense polarisation. For example, during the 25 January protests in 2016, more than 20,000 tweets were shared in Egypt, one hashtag #Iparticipated_in_January_Revolution became particularly trending. The hashtag was created after the arrest of the Egyptian doctor Taher Mukhtar, who was accused of preparing and participating in the protests on 25 January 2016. People began to use the hashtag to oppose the idea of the revolution becoming a crime and this led to an online battle between the Egyptian revolutionaries and those who opposed it with the hashtag I_didn't_participate_in_January_setback (Gamal and Inwood 2016). These types of counter-movements on the Internet induce the dissidents to engage in the old fights again and again.

Yet, the digitally supported protests did not completely fail to bring change to Arab societies. Although democratic societal change is still ongoing in these countries, it is possible to see change in the way people conceive concepts such as power and the state. The interviews for this research indicated that the 2011 protests in Egypt brought about psychological change to the way people engage with dominant powers such as the state. Sami,[3] for instance, defines this change as follows:

> It is going to be decades until you can see things get stable. The most important change that happened during the January 25 protests was in the psychology of the people. The protests gave people the idea that things can change and that they can do things in the name of the Egyptian people. Right now, people are taking the revolution in the wrong direction, but it is still the people who decide which way we should go.

The digitally supported protests introduced a culture of rebelling and compelling to these Arab societies. This ran in parallel with the growing number of Internet users in both countries since 2011. In Egypt, the number of Internet users was 49.23 per cent by 2018, with active social media users at 36 million (Reda 2018). In Syria, the number of citizen journalists soared from the beginning of the uprising. There was also a sharp increase in the number of Facebook pages, opposition web pages and social media groups (Harkin 2013). Recent years also saw the emergence of Internet projects like the Hawa Net project, which was the first wireless project that aims to cover new places with wireless towers (Al Khatieb 2015). The rise of cyber armies and the effective Internet users in Syria and Egypt indicates the conditions that are present for more digitally supported battles between oppressive regimes and dissidents in the upcoming years in the Arab region. However, as long as the spirit of change continues to be articulated in both the online and offline spheres and the public actively engage with it, ultimately the invisibility of repressive and corrupt regimes will be broken and social change in Arab societies will become ever more inevitable.

Notes

1. Interview with Wael Abbas, Egyptian blogger, Cairo, 2014.
2. Interview with Abdelrahman Mansour, administrator of the Kullena Khaled Said Facebook page, London, 2015.
3. Interview with Sami (pseudonym), Egyptian activist, Cairo, 2014.

REFERENCES

Al Khatieb, M. (2015, April 14). Seeking Internet access, Syrians turn to Turkey's wireless network. *Al-Monitor*. Retrieved from http://www.al-monitor.com/pulse/originals/2015/04/aleppo-rebel-control-internet-networks-syria-turkey.html#.

Angelis, E. (2011). The state of disarray of a networked revolution. *Sociologica, 3*, 1–24. https://doi.org/10.2383/36423.

Aouragh, M., & Alexander, A. (2011). The Arab Spring: The Egyptian experience: Sense and nonsense of the Internet revolution. *International Journal of Communication, 5*, 1344–1358.

Bayat, A. (2013). *Life as politics: How ordinary people change the Middle East.* California, CA: Stanford University Press.

Bennett, W. L., & Segerberg, A. (2013). *The logic of connective action.* Cambridge, UK: Cambridge University Press.

Castells, M. (2012). *Networks of outrage and hope: Social movements in the Internet age.* Cambridge, UK: Polity Press.

Cole, J. (2014). *The new Arabs: How the millennial generation is changing the Middle East.* New York, NY: Simon & Schuster.

El-Sherif, A. (2014, January 29). *Egypt's post-Mubarak predicament.* Carnegie Endowment for International Peace. Retrieved from http://carnegieendowment.org/2014/01/29/egypt-s-post-mubarak-predicament.

Eltantawy, N., & Wiest, J. (2011). Social media in the Egyptian revolution: Reconsidering resource mobilization theory. *International Journal of Communication, 5*, 1207–1224.

Gamal, R., & Inwood, J. (2016, January 25). *Egypt's revolution five-year anniversary sees hashtag battle.* BBC. Retrieved from https://www.bbc.co.uk/news/av/world-middle-east-35384770/egypt-s-revolution-five-year-anniversary-sees-hashtag-battle.

Gerbaudo, P. (2012). *Tweets and the streets: Social media and contemporary activism.* New York, NY: Pluto Press.

Harkin, J. (2013, April). Is it possible to understand the Syrian revolution through the prism of social media? *Westminster Papers in Communication and Culture, 9*(2), 94–112.

Hinnebusch, R. (2012). Syria: From 'authoritarian upgrading' to revolution? *International Affairs, 88*(1), 95–113. https://doi.org/10.1111/j.1468-2346.2012.01059.x.

Howard, P., & Hussain, M. (2013). *Democracy's fourth wave? Digital media and the Arab Spring.* New York, NY: Oxford University Press.

Howard, P., & Parks, M. (2012). Social media and political change: Capacity, constraint, and consequence. *Journal of Communication, 62*(2), 359–362. https://doi.org/10.1111/j.1460-2466.2012.01626.x.

Human Rights Watch. (2018a, November 18). *Egypt: Mass arrests of lawyers, activists*. Human Rights Watch. Retrieved from https://www.hrw.org/news/2018/11/18/egypt-mass-arrests-lawyers-activists.

Human Rights Watch. (2018b, June 29). *Syria: Events of 2018*. Human Rights Watch. Retrieved from https://www.hrw.org/world-report/2019/country-chapters/syria.

Jenkins, J. C. (1983). Resource mobilisation theory and the study of social movements. *Annual Review of Sociology, 9,* 527–553.

Landis, J. (2011, March 31). *Speech to the Syrian parliament by president Bashar Al-Assad: Wednesday, March 30, 2011* [Web log post]. Retrieved from http://www.joshualandis.com/blog/speech-to-the-syrian-parliament-by-president-bashar-al-assad-wednesday-march-30-2011/.

Magout, M. (2012, June). *Cultural dynamics in the Syrian uprising*. Graduate Section of the British Society for Middle Eastern Studies Annual Conference. London: The London School of Economics and Political Science.

McCarthy, J. D., & Zald, M. N. (1977). Resource mobilization and social movements: A partial theory. *American Journal of Sociology, 82*(6), 1212–1241.

Morris, A., & Staggenborg, S. (2002, November). *Leadership in social movements*. Retrieved from https://www.sociology.northwestern.edu/documents/faculty-docs/faculty-research-article/Morris-Leadership.pdf.

Reda, L. (2018, May 30). Analysis: What are Egyptians using the Internet for? *Egypt Today*. Retrieved from http://www.egypttoday.com/Article/3/50919/Analysis-What-are-Egyptians-using-the-internet-for.

Rinke, E. M., & Roder, M. (2011). Media ecologies, communication culture and temporal spatial unfolding: Three components in a communication model of the Egyptian regime change. *International Journal of Communication, 5*(13), 1273–1285.

Stacher, J. (2012). *Adaptable autocrats: Regime power in Egypt and Syria*. Stanford, CA: Stanford University Press.

Tarrow, S. (2011). *Power in movement: Social movements and contentious politics*. New York, NY: Cambridge University Press.

Tilly, C. (2006). *Regimes and repertoires*. Chicago, IL: University of Chicago Press.

Trere, E. (2019). *Hybrid media activism: Ecologies, imaginaries, algorithms*. New York, NY: Routledge.

Tufekci, Z. (2017). *Twitter and tear gas: The power and fragility of networked protest*. New Haven, CT: Yale University Press.

Van Dam, N. (1996). *The struggle for power in Syria: Politics and society under Assad and the Ba'ath Party*. New York, NY: I.B. Tauris.

Weeden, L. (1999). *Ambiguities of domination: Politics, rhetoric, and symbols in contemporary Syria*. Chicago, IL: University of Chicago Press.

We will not be silenced: April 6, after court order banning group. (2014, April 28). *Ahram Online*. Retrieved from http://english.ahram.org.eg/NewsContent/1/64/100015/Egypt/Politics-/We-will-not-be-silenced-April-,-after-court-order-.aspx.

Index